ALSO FROM VISIBLE INK PRESS

The Handy Science Answer Book, 2nd edition

Can any bird fly upside down? Is white gold really gold? Compiled from the ready-reference files of the Science and Technology Department of the Carnegie Library of Pittsburgh, this best seller answers 1,400 questions about the inner workings of the human body and outer space, about math and computers, and about planes, trains, and automobiles. By the Science and Technology Department of the Carnegie Library of Pittsburgh, 7.25" x 9.25", 598 pages, 100 illustrations, dozens of tables, $16.95, ISBN 0-7876-1013-5.

The Handy Weather Answer Book

What's the difference between sleet and freezing rain? Do mobile homes attract tornadoes? What exactly is wind chill and how is it figured out? How can the temperature be determined from the frequency of cricket chirps? You'll find clear-cut answers to these and more than 1,000 other frequently asked questions in *The Handy Weather Answer Book*. By Walter A. Lyons, 7.25" x 9.25", 398 pages, 75 illustrations, $16.95, ISBN 0-7876-1034-8.

The Handy Space Answer Book

Is there life on Mars? Did an asteroid cause the extinction of dinosaurs? Find answers to these and 1,200 other questions in *The Handy Space Answer Book*. It tackles hundreds of technical concepts—quasars, black holes, NASA missions and the possibility of alien life—in everday language. With vivid photos, thorough indexing and an appealing format, it's as fun to read as it is informative for space lovers and curious readers of all ages. By Phillis Engelbert and Diane L. Dupuis, 7.25" x 9.25", 590 pages, $17.95, ISBN 1-57859-017-5.

The Handy Bug Answer Book

For anyone who's asked the question, "Why is the world full of bugs?" and for the child in the rest of us, there's *The Handy Bug Answer Book*. Author Dr. Gilbert Waldbauer shares his enthusiasm about the intricate and barely visible world of insects in this entertaining and accessible question-and-answer book. Organized by highly browsable topics, *Handy Bug* answers nearly 800 questions on insect lives and habits, including their number, sex lives, physical makeup, where they can be found, which are pests and which are good to have around, and the differences between insects, spiders, millipedes, and other invertebrates. By Dr. Gilbert Waldbauer, 7.25" x 9.25", 308 pages, 70 illustrations, including 37 color photos, $19.95, ISBN 1-57859-049-3.

The Handy Earth Answer Book

The Earth is the world's biggest celebrity. Natural disasters. . .weather. . .global warming. . .and man's curious relationship with the highest mountains, deepest oceans, and extremes of heat and cold are always in the news. To satisfy everyone's Earthly desires, there's *The Handy Earth Answer Book*. Filled with plain-language answers to 1,000 commonly asked questions, *Handy Earth* is an easy-to-use resource with broad appeal. Parents, students, and armchair scientists will enjoy this highly browsable reference, which covers physical geology, mountains, plate tectonics, geophysics, geochemistry, oceanography, and other subjects in straightforward, question-and-answer format. By Dr. John Ernissee and Dr. Frank Vento, 7.25" x 9.25", 400 pages, 100 illustrations, including 40 color photos, $19.95, ISBN 1-57859-0493.

The Handy Geography Answer Book

As the most up-to-date geographic reference available, *The Handy Geography Answer Book* offers nontechnical explanations of the latest science with maps and information on nations and states. Skillfully combining entertainment and informtaion, it covers common trivia questions—highest, tallest, deepest, hottest, etc.—but also explores the social and political landscape, ranging from the influence of geography on language, religion, and architecture to how migration, population, and locations of countries and cities has been affected by terrain. By Matt T. Rosenberg, 7.25" x 9.25", 435 pages, 110 illustrations, including 30 color maps and photos, $19.95, ISBN 1-57859-062-0.

The Handy Sports Answer Book

What makes a curveball curve? What was the longest basketball winning streak of all time? What Supreme Court justice once won the NFL rushing title? When did professional baseball begin? Despite an endless stream of encyclopedias and single-sport references, few, if any, speak to the most commonly asked questions surrounding America's favorite sports. *The Handy Sports Answer Book* fills that gap by providing an easy-to-use, entertaining, and broadly based reference that sports fans will use to expand their knowledge, extend their interests, settle arguments, and hone their trivia skills. By Kevin Hillstrom, Laurie Hillstrom, and Roger Matuz, 7.25" x 9.25", 446 pages, 150 illustrations, including 35 color photos, $19.95, ISBN 1-57859-075-2.

THE
HANDY
GEOGRAPHY
ANSWER
BOOK

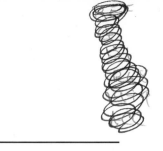

THE
HANDY
GEOGRAPHY
ANSWER
BOOK™

MATTHEW T. ROSENBERG

VISIBLE
INK
PRESS

DETROIT · LONDON

The Handy Geography Answer Book™

COPYRIGHT © 1999 BY VISIBLE INK PRESS

Published by Visible Ink Press™
a division of Gale Research
27500 Drake Rd.
Farmington Hills, MI 48331-3535

Visible Ink Press and The Handy Geography Answer Book are trademarks of Gale Research.

Most Visible Ink Press™ books are available at special quantity discounts when purchased in bulk by corporations, organizations, or groups. Customized printings, special imprints, messages, and excerpts can be produced to meet your needs. For more information, contact Special Markets Manager, Gale Research, 835 Penobscot Bldg., Detroit, MI 48226-4094.

Art Director: Cindy Baldwin
Typesetting: The Graphix Group

Country information in Chapter 21 was compiled from the C.I.A. World Factbook and from the U.S. Census Bureau.

Library of Congress Cataloging-in-Publication Data

Rosenberg, Matthew Todd, 1973–
    The handy geography answer book / Matthew Todd Rosenberg.
        p. cm.
    Includes bibliographical references and index.
    ISBN 1-57859-062-0 (softcover)
        1. Geography—Miscellanea. I. Title
    G131.R68   1999
    910'.76—dc21
                                                    98-36290
                                                        CIP

About the Author

Having long been intrigued by maps and foreign lands, Matthew Todd Rosenberg became a geography major in his freshman year at the University of California, Davis, where he graduated with a bachelor's degree in geography. He is currently a graduate student at California State University, Northridge, and lives in Santa Clarita, California. Matt works for the Mining Co. internet complex where he operates a very popular geography web page (http://geography.miningco.com). At the site, he writes weekly articles about geography, answers hundreds of e-mail questions each month, creates and distributes a weekly e-mail newsletter, and moderates a chat room and bulletin board devoted to geography.

Matt is a former Coordinator of Disaster Services for the American Red Cross and has assisted victims of disaster across the United States, including those of Hurricane Andrew in 1992, the Midwest Floods of 1993, and the Northridge Earthquake in 1994.

A Ph.D. candidate, Matt is a member of the Association of American Geographers and the National Council for Geography Education. Matt enjoys hearing from readers, so visit his web site and contact him there.

Contents

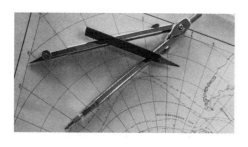

DEFINING THE WORLD . . . 1

MAPS . . . 19

PHYSICAL ENVIRONMENT . . . 37

WATER AND ICE . . . 49

Oceans and Seas . . . Rivers and Lakes . . . Precipitation . . . Glaciers . . . Controlling Water

CLIMATE . . . 69

Definitions . . . The Atmosphere . . . Ozone . . . Climatic Trends . . . Weather . . . Wind

HAZARDS . . . 83

Volcanoes . . . Earthquakes . . . Tsumanis . . . Hurricanes . . . Floods . . . Tornadoes . . . Other Hazards and Disasters

TRANSPORTATION AND URBAN GEOGRAPHY . . . 103

Urban Sprawl . . . Cities and Suburbs . . . Urban Structures . . . Transportation . . . Flight . . . Roads and Railways . . . Sea Transport

POLITICAL GEOGRAPHY . . . 117

The United Nations . . . NATO and the Cold War . . . The World Today . . . Colonies and Expansionism . . . The World Economy

CULTURAL GEOGRAPHY . . . 131

Population . . . Language and Religion . . . Dealing with Hazards . . . Culture around the World

TIME, SEASONS, AND CALENDARS . . . 143

Time . . . Time Zones . . . Daylight Savings Time . . . Keeping Time . . . Calendars . . . The Seasons

EXPLORATION . . . 157

Europe and Asia . . . Africa . . . The New World . . . The Poles

UNITED STATES OF AMERICA . . . 169

Physical Features and Resources . . . The States . . . Cities and Counties . . . People and Culture . . . History: Creating the United States of America

NORTH AMERICA . . . 191

Greenland and the North Pole Region . . . Canada . . . Mexico . . . Central Ameria . . . The West Indies

AFRICA . . . 279

Physical Features and Resources . . . History . . .
People, Countries, and Cities

OCEANIA AND ANTARCTICA . . . 293

Oceania . . . Australia . . . New Zealand . . .
Antarctica

COUNTRIES OF THE WORLD . . . 305

Key geographic, political, and cultural information on the 192 countries of the World

Introduction

Since I was a very young boy, I have made dams out of mud, climbed rocks to find what was just over the next hill, and learnd about other places around the world. Maybe that's where I came to my lifelong fascination with geography. I find geography to be a profoundly stimulating field—one that I'm happy to share in this fun, easy-to-browse and adventurous format.

Though many consider geography soley to be limited to names of countries and their capitals, the science is really a synthesis of all fields, actions, events, and places that deal with humans and the Earth. Geography tells you why certain places are called what they are, how the boundaries are defined, the cultural characteristics of the place, and how all of these factors affect the local inhabitants or extend to the world at large. Why was Germany divided in two, and why was it reunited? How did Israel become a country? Where is Timbuktu?

Presented at this level, geography is an exciting science that not only helps you define boundaries, capitals, and cultures, but gives you a better grasp of today's world and current events. *The Handy Geography Answer Book* will improve your understanding of the nightly news, providing perspective on the headlines and hot topics. How many countries are there in Africa? Where is the outback in Australia? Where does Asia end and Europe begin? How big is Russia? How many islands make up Indonesia? What is the Middle East in the middle of?

The Handy Geography Answer Book also delivers a sense of wonder of our world. What is the North Magnetic Pole? What is the highest point in the world? How fast does the Earth spin? What is the largest landlocked country in the world? What is the world's highest city? What are the seven seas? What is the difference between a bay and a gulf? How does land turn into desert? What is the most populated urban area in the world? Where is the world's tallest waterfall?

You'll find answers to these and 1,000 other geographical questions in the 21 chapters of *The Handy Geography Answer Book.* Chapters 1 through 11 cover geographical themes; chapters 12 through 20 focus on world regions; and chapter 21 includes fascinating data about each and every country of the world. In addition, 110 photographs and maps help bring places around the world to life.

I hope that I have succeeded in presenting geography in the exciting and dynamic light in which I see it, and that you will enjoy the give and take in the *The Handy Geography Answer Book*. The "mother of all sciences" is a fascinating, endlessly intriguing quest for knowledge.

Acknowledgments

First and foremost, I would like to thank my dear wife and best friend, Jennifer Rosenberg, for her support, selflessness, and wisdom in helping me with this book. It was truly a partnership and I am extremely and profoundly grateful, for without her help, this book would surely not exist.

I would like to thank Dr. Dennis Dingemans, a professor who throughout my undergraduate years at University of California, Davis, molded my enthusiasm for geography into a viable academic career. I would also like to thank all of my professors at California State University, Northridge, especially Dr. Darrick Danta, Dr. James Allen, and Dr. William Bowen, for their support and enthusiasm.

My family has been an important source of inspiration and strength, supporting me during this momentous project. I owe a debt of gratitude to my mom, Barbara Rosenberg, who gave me self esteem and granted me space to explore; to my in-laws, Kenneth and Leslie Thornton, for their support and guidance; to my grandfather, Abbott Rosenberg, for the multitude of questions and ideas that he proffered; to my sister, Laura, for the cookies that kept me going; and to the rest of my family for their love and patience.

I owe a huge 'thank you' to my editor at Visible Ink Press, Jeffrey Hermann, for all of his assistance. Additional credit goes to Christa Brelin and Julia Furtaw for their guidance and support, Bob Huffman for his copywriting efforts, Cindy Baldwin and Michelle DiMercurio for their artistic expertise, Pam Reed, Randy Bassett, and Gary Leach for their help with the images in this book, Jeff Muhr for everything technical, Matt Nowinski for copyediting assistance, Diane Dupuis for proofreading, Heather Mack for her help with image cataloging, Barbara Cohen for the indexing, and Marco Di Vita of the Graphix Group for his typesetting skills. I would also like to thank David Deis at Dreamline Cartography for the great maps that are included in this book.

Finally, I would like to thank you, the reader, for reading my book.

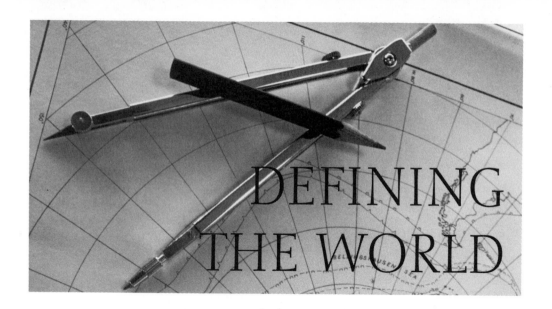

DEFINING THE WORLD

What does the word **"geography"** mean?

The word geography is of Greek origin and can be divided into two parts, *geo,* meaning the Earth, and *graphy,* which refers to writing. So geography can be loosely translated to "writing about the Earth." Ancient geography was often descriptions of far away places, but modern geography has become much more than writing about the Earth. Contemporary geographers have a difficult time defining the discipline. Some of my favorite definitions include "the bridge between the human and the natural sciences," "the mother of all sciences," and "anything that can be mapped."

Who **invented** geography?

The Greek scholar Eratosthenes is credited with the first use of the word geography in the third century B.C.E. He is also known as the "father of geography" for his geographical writing and accomplishments, including the measurement of the circumference of the Earth.

What are **continents**?

Continents are the six or seven large land masses on the planet. If you count seven continents these include Europe, Asia, Africa, Australia, Antarctica, North America, and South America. Some geographers refer to six continents by combining Europe and Asia as Eurasia, due to the

1

fact that it is one large tectonic plate and land mass. So whether you count Europe and Asia one continent or two (divided at the Ural Mountains in western Russia) is up to the individual. Australia is the only continent that is its own country.

What is a **subcontinent?**

A subcontinent is a landmass that has its own continental shelf and its own continental plate. Currently, India and its neighbors form the only subcontinent, but in millions of years, Eastern Africa will break off from Africa and become its own subcontinent.

What was **Pangea**?

About 250 million years ago, all of the land on Earth was lumped together into one large continent known as Pangea. Faults and rifts broke the land masses apart and pushed them away from each other. The continents slowly moved across the Earth to their present positions, and they continue to move today. The Indian subcontinent (composed of India and its neighbors) continues to push into Asia and create the Himalayas.

What is the **circumference** of the Earth?

The circumference of the Earth at the equator is 24,901.55 miles. Due to the irregular, ellipsoid shape of the Earth, a line of longitude wrapped around the Earth going through the north and south poles is 24,859.82 miles. Therefore, the Earth is a little bit (about 41 miles) wider than it is high. The diameter of the Earth 7,926.41 miles.

Is the Earth a **perfect sphere**?

No, the Earth is a bit wider than it is "high." The shape is often called a geoid (Earth-like) or an ellipsoid. The rotation of the Earth causes a slight bulge towards the equator. The circumference of the Earth at the equator (24,901.55) is about 41 miles greater than the circumference through the poles (24,859.82 miles). If you were standing on the moon,

When did geography begin?

We must assume that one of the earliest questions human beings asked was "What's over that hill?" Geographic thought has been present for thousands of years—maps drawn in the sand or etched in stone, as well as explorations to distant lands, were made by the earliest civilizations. Geographic knowledge has been accumulating since the beginning of humankind.

looking back home, it would be virtually impossible to see the bulge and the Earth would appear to be a perfect sphere (which it practically is).

If the Earth is so large, why did **Colombus think that India was close** enough to reach by sailing west from Europe?

The Greek geographer Posidonus did not believe Eratosthenes' earlier calculation, so he performed his own measurement of the Earth's circumference and arrived at the figure of 18,000 miles. Columbus used the circumference estimated by Posidonus when he argued his plan before the Spanish court. The 7,000 mile difference between the actual circumference and the one Columbus used led him to believe he could reach India rather quickly by sailing west from Europe.

What is the **North Magnetic Pole**?

The North Magnetic Pole is where compass needles around the world point. It is located in Canada's Northwest Territories at about 71 degrees north, 96 degrees west (latitude and longitude), about 900 miles away from absolute North Pole. It moves continuously, so to determine true north, look at a recent topographic map for your local area. It should note the "magnetic declination," which means the degrees east or west that you'll need to rotate your compass to determine which way is actually north.

How was the circumference of the Earth determined?

The Greek geographer and librarian at the Great Library of Alexandria, Eratosthenes (c.273–c.192 B.C.E.), was aware that the sun reached the bottom of a well in Egypt only once a year, on the first day of summer. The well was near Aswan and the Tropic of Cancer (where the sun is directly overhead at noon on the summer solstice). Eratosthenes estimated the distance between the well and Alexandria based on the length of time it took camel caravans to travel between the two places. He measured the angle of the sun's shadow in Alexandria at the same time as the well was lit by the sun, and then used a mathematic formula to determine that the circumference of the Earth was 25,000 miles—amazingly close to the actual figure!

Have compasses always pointed **north**?

No, they have not. Though compasses always point to the magnetic pole, the magnetic pole has not always been in the north. Every 300,000 to 1 million years, the magnetic pole flips from north to south or from south to north. If compasses had been around before the last time the magnetic pole reversed, their arrows would have pointed south rather than north.

How, why, and how much does **Magnetic North move**?

Scientists aren't sure why the Earth's magnetic pole moves, only that it does. The amount of movement varies, but it's never more than a few miles each year.

What is an **azimuth**?

Azimuth is another method for stating compass direction. It is based on the compass as 360 degrees, with north at 0 degrees, east at 90, south at

180, and west at 270 degrees. You can refer to a direction as "head 90 degrees" instead of "head east."

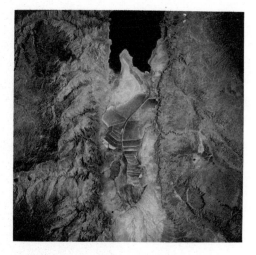

A portion of the Dead Sea in Jordan from a satellite image. (NASA/Corbis)

What is the **lowest point** in the world on land?

The world's lowest point is at the Dead Sea on the border of Israel and Jordan. It is 1,312 feet below sea level.

What are the **lowest points** on each continent?

In Africa, the lowest point is Lake Assal in Djibouti, 512 feet below sea level. In North America, California's Death Valley lies at 282 feet below sea level. Argentina's Bahia Blanca is the lowest point in South America at 138 feet below sea level. The Caspian Sea in Europe lies at 92 feet below sea level, and Australia's lowest point is a mere 52 feet below sea level at Lake Eyre.

What is the **highest point** in the world?

The highest point above sea level in the world is Mount Everest, which lies on the border of China and Nepal, at 29,082 feet.

What are the **highest points** on each continent?

The highest peak in South America, Aconcagua, lies in Argentina at 22,834 feet. In North America, Alaska's Mt. McKinley (also called Denali, as it is known indigenously) is 20,320 feet. The famous Mt. Kilimanjaro (19,340 feet) is in Africa's Tanzania. Ice-covered Antarctica's high point is known as Vinson Massif, 16,864 feet. Europe's Mont Blanc

5

is in the Alps between France and Italy at 15,771 feet. Australia's high point, Kosciusko, is the lowest of all the continents at 7,310 feet.

What is the **deepest point** in the ocean?

The western Pacific Ocean's Marianas Trench is the deepest point in the world at 35,840 feet (almost seven miles) below sea level.

What are the **deepest points** in the oceans?

In the Atlantic Ocean, the Puerto Rico Trench is 28,374 feet below the surface. In the Arctic Ocean, the Eurasia Basin is 17,881 feet deep. The Java Trench in the Indian Ocean is 23,376 feet deep.

Where is the **farthest point from land**?

In the middle of the Southern Pacific ocean lies a spot that is 1,600 miles from any land. Located at 47 degrees, 30' south, 120 degrees west, this spot is equidistant from Antarctica, Australia, and Pitcairn Island.

Where is the **farthest point from an ocean**?

In northern China lies a spot that is over 1,600 miles from any ocean. Located at 46 degrees, 17' north, 86 degrees, 40' east, the land is equidistant from the Arctic Ocean, Indian Ocean, and Pacific Ocean.

What is a **hemisphere**?

A hemisphere is half of the Earth. The Earth can actually be divided into hemispheres in two ways: by the equator, and by the Prime Meridian (through Greenwich, England) at zero degrees longitude and another meridian at 180 degrees longitude (near the location of the International Dateline in the western Pacific Ocean). The equator divides the Earth into northern and southern hemispheres. There are seasonal differences between the northern and southern hemispheres but there is no such

Mount Everest, the highest point in the world. (Archive Photos)

difference between the eastern and western hemispheres. Zero and 180 degrees longitude divide the Earth into the eastern (most of Europe, Africa, Australia, and Asia) and western (the Americas) hemispheres.

How **fast does the Earth spin**?

It depends on where you are on the planet. If you were standing on the north pole or close to it, you would be moving at a very slow rate of speed—nearly zero miles per hour. On the other hand, those who live at the equator (and therefore have to move about 24,900 miles in a 24 hour period), zoom at about 1,038 miles per hour. Those in the mid-latitudes, as in the United States, breeze along from about 700 to 900 miles per hour.

Why **don't we feel** the Earth moving?

Even though we constantly move at a high rate of speed, we don't feel it, just as we don't feel the speed at which we're flying in an airplane or dri-

What is the world's largest island?

The world's largest island is Greenland, technically now known as Kalaallit Nunaat. Greenland is located in the North Atlantic Ocean near Canada. It is a territory of Denmark but has locally governed itself since 1979. It is approximately 840,000 square miles. Australia, while it also meets the usual definition of an island (surrounded by water) and is larger than Greenland, is not considered an island but a continent.

ving in a car. It's only when there is a sudden change in speed that we notice, and if the Earth made such a change we would certainly feel it.

Does the Earth **spin at a constant** rate?

The rotation of the Earth actually has slight variations. Motion and activity within the Earth, such as friction due to tides, wind, and other forces, change the speed of the planet's rotation a little. These changes only amount to milliseconds over hundreds of years but do cause people who keep exact time to make corrections every few years.

What is the **axis** of the Earth?

The axis is the imaginary line that passes through the north and south poles about which the Earth revolves.

What is **inside the Earth**?

At the very center of the Earth is a dense and solid inner core of iron and other minerals that is about 1,800 miles wide. Surrounding the inner core is a liquid (molten) outer core. Surrounding the outer core is the mantle, which makes up the bulk of the interior of the Earth. The mantle is composed of three layers—two outer layers are solid and the inner layer (the asthenosphere) is a layer of rock that is easily moved and shaped.

What is the **largest continent**?

The largest continent is Eurasia (Europe and Asia combined) at 21,100,000 square miles. But even if you consider Europe and Asia to be two separate continents, Asia is still the largest, at 17,300,000 square miles.

Why is **Greenland considered an island** while **Australia is a continent**?

Australia is three and a half times larger than Greenland and comprises most of the land on the Indo-Australian plate, while Greenland is distinctly part of the North American plate.

What island did **Robinson Crusoe** shipwreck on?

Daniel Defoe based his novel *Robinson Crusoe* on the story of Alexander Selkirk. Selkirk was an English sailor who had an argument with the captain of his ship and asked to go ashore on the island of Mas a Tierra (also known as Robinson Crusoe Island), about 400 miles west of Chile. Selkirk was stranded on the island from 1704 to 1709, when he was rescued by another English ship.

If I **dug through the Earth**, would I end up in China?

If you are in North America and you were able to dig through the Earth (which is impossible due to such things as pressure, the molten outer core, and solid inner core), you would end up in the Indian Ocean, far from land masses. If you were really lucky, you might end up on a tiny island, but you're surely not going to end up in China. The points at opposite sides of the Earth are called antipodes. Most antipodes of Europe fall into the Pacific Ocean.

The Hanging Gardens of Babylon, one of the seven wonders of the ancient world. (Archive Photos)

What were the **seven wonders of the ancient world**?

While there was often disagreement by ancient and classical scholars as to which major works of art and architecture could be considered wonders, these seven were nearly always on the list: the Pyramids of Egypt (the only remaining wonder), the Colossus of Rhodes (a very large bronze statue), The Temple of Artemis at Ephesus (a marble temple in Turkey), The Mausoleum at Halicarnassus (a tomb in Turkey), The Statue of Zeus at Olympia (an ivory and gold statue), The Hanging Gardens of Babylon (a brick terrace with hundreds of plants and flowers), and The Lighthouse of Alexandria.

What are the **seven wonders of the modern world**?

According to the American Society of Civil Engineers, the seven wonders of the modern world include the Channel Tunnel between England and France; the CN Tower in Toronto, Canada; the Empire State Building, New York; the Golden Gate Bridge, San Francisco; the Itaipu Dam between Brazil and Paraguay; the Netherlands North Sea Protection Works; and the Panama Canal.

The CN Tower in Toronto, Canada, one of the seven modern wonders. (Michael S. Yamashita/Corbis)

What are the **seven natural wonders of the world**?

These include the Aurora Borealis (northern lights), Mount Everest, Victoria Falls (in eastern Africa), the Grand Canyon (USA), Great Barrier Reef (Australia), Paricutin (volcano in Mexico), and the harbor of Rio de Janeiro (Brazil) with its stunning topography.

What is an **archipelago**?

An archipelago is a chain (or group) of islands that are close to one another. The Aleutian Islands of Alaska and the Hawaiian Islands are both archipelagos. They are usually formed by plates pushing into one another or by volcanic activity.

What are the **Arctic and Antarctic Circles**?

The circles are imaginary lines that surround the north and south poles at 66.5 degrees latitude. The Arctic Circle is a line of latitude at 66.5 degrees north of the equator and the Antarctic Circle is a line of latitude

What is a strait?

A strait is a narrow body of water between islands or continents that connects two larger bodies of water. Two of the most famous straits are the Strait of Gibraltar, which connects the Mediterranean Sea and the Atlantic Ocean, and the Strait of Hormuz, which connects the Persian Gulf to the Indian Ocean between Yemen and Dijbouti.

at 66.5 degrees south. Areas north of the Arctic Circle are dark for 24 hours near December 21 and areas south of the Antarctic Circle are dark for 24 hours near June 21. Almost all of the continent of Antarctica is located to the south of the Antarctic circle.

What is **subduction**?

When two tectonic plates meet and collide, crust must either be lifted up, as in the case of the Himalayas, or it must be sent back into the Earth. When crust from one plate slides under the crust of another, it is called subduction, and the area around the subduction is called a subduction zone.

What is the **mid–Atlantic ridge**?

We don't get to appreciate the beauty of this huge mountain range because it's located at the bottom of the Atlantic Ocean (with one exception: Iceland is a part of the ridge). The ridge is a crack between tectonic plates where new ocean floor is being created as magma flows up from under the Earth. As more crust is created, it pushes the older crust further away. The new crust at the ridge piles up to form mountains and then begins to move across the bottom of the ocean. Because the Earth can't get larger as more crust is created, the crust eventually has nowhere to go except back into the Earth. This is where subduction occurs.

When did **agriculture** begin?

Agriculture began about 10 to 12 thousand years ago in a time period known as the first agricultural revolution. It was at this time that humans began to domesticate plants and animals for food. Before the agricultural revolution, people relied on hunting wild animals and gathering wild plants for nutrition. This revolution took place almost simultaneously in different areas of human settlement.

When was the second **agricultural revolution**?

The second agricultural revolution occurred in the 17th century. During this time, production and distribution of agricultural products were improved through machinery, vehicles, and tools, which allowed more people to move away from the farm and into the cities. This mass migration from rural areas to urban areas coincided with the beginning of the industrial revolution.

What was the **industrial revolution**?

The industrial revolution began in the 18th century in England with the transformation from an agricultural-based economy to an industrial-based economy. It was a period of increased development in industry and mechanization that improved manufacturing and agricultural processes, thereby allowing more people to move to the cities. It included the development of the steam engine and the railroad.

What is the **green revolution**?

The green revolution began in the 1960s as an effort by international organizations (especially the United Nations) to help increase the agricultural production of less developed nations. Since that time, technology has helped improve crop output, which is reaching all-time highs throughout the world.

Which country has the most neighbors?

China is bordered by 13 neighbors: Mongolia, Russia, North Korea, Vietnam, Laos, Myanmar, India, Bhutan, Nepal, Afghanistan, Tajikistan, Kyrgyzstan, and Kazakhstan. Russia is next in line, as it shares its border with 12 other countries. Brazil is third with nine neighbors.

How much of the world's population is **devoted to agriculture**?

In less-developed countries, such as Asia and Africa, a majority of the population in engaged in agricultural activity. In the more-developed countries of Western Europe and North America, less than one tenth of the population relies on agriculture for their livelihood.

How were **animals first domesticated**?

Dogs were probably some of the first animals to become domesticated. Wild dogs probably came close to human villages scavenging for food and were quickly trained as companions and protectors. Over time, early agriculturalists realized the value of domesticating other animals and proceeded to do so. Many different kinds of animals were domesticated in different areas of the world.

What is the **largest country in the world**?

Russia is by far the largest at about 6.6 million square miles. Russia is followed in size by Canada, China, the United States, Brazil, Australia, India, Argentina, Kazakhstan, and Sudan.

Which country has the **most people**?

China is the most populous country in the world with 1.2 billion people. That means that about one out of every five people on the planet is Chinese. India is the second most populous country and is home to 970 million people, the United States is third with 267 million, Indonesia is fourth with 204 million, and Brazil is fifth with 160 million.

Which country has the **longest coast line**?

The coastline of Canada and its associated islands is the longest in the world, about 151,400 miles long. Russia, which is the largest country in the world, has the second longest coastline at about 23,400 miles.

Which countries **have the fewest neighbors**?

All island nations (such as Australia, New Zealand, Madagascar, etc.) have no neighbors. Hatiti, Dominican Republic, Papua New Guinea, Ireland, UK, and many others are island nations that share an island.

Which **non-island nations have the fewest neighbors**?

There are 10 non-island countries that share a land border with just one neighbor: Canada (neighboring USA), Monaco (France), San Marino (Italy), Vatican (Italy), Qatar (Saudi Arabia), Portugal (Spain), Gambia (Senegal), Denmark (Germany), Lesotho (South Africa), and South Korea (North Korea).

How does a city get chosen to **host the Olympics**?

The International Olympic Committee chooses a city as an Olympic site through a complex process. Cities (and their countries) are judged on many characteristics, including environmental protection, climate, security, medical services, immigration, housing, and many others. Cities eagerly spend millions of dollars in construction and preparation for possible selection as a host city, as an investment in the city's future.

How is a **capital different from a capitol**?

The capital is a city and the capitol is a building. The capitol is located in the capital. To remember the difference, think about the "o" in capitol as being the dome of a capitol building. Capital cities are often the largest cities in a country or region.

What did the average European know about the world in the **Middle Ages**?

In Europe in the Middle Ages, most individuals' knowledge of the world was quite limited. Geographic knowledge developed by the Greeks and Romans (who knew the Earth was a sphere) was all but lost in Europe. Europeans of the time thought of the world as flat and composed of only Europe, Asia, and Africa.

Where is the **third world**?

Originally, the third world referred to those countries that did not align themselves with the United States (first world) or the Soviet Union (second world) during the cold war. Over time, the term took on different meanings and has come to refer to less-developed or developing nations, which are the more preferred terms.

What is the **largest landlocked country** in the world?

Kazakhstan, which is the ninth largest country in the world, has no outlet to the ocean. It is over one million square miles in area. While Kazakhstan is located adjacent to the Caspian Sea, the Caspian Sea is a landlocked sea.

Where are **cyberspace and the Internet**?

Cyberspace is not space in the old-fashioned sense of the word at all. The Internet is composed of thousands of computers around the world, which are connected to each other in order to provide information across cyberspace as though there were no global boundaries, mountain

Cartographers map neigborhoods, cities, states, countries, the world, and even other planets. (The National Archives/Corbis)

passes, or oceans to cross. When you send e-mail to a friend on the other side of the planet, it passes from your computer to that of your Internet service provider and then from computer to computer, making its way to your friend in a matter of seconds. Similarly, when you access a page on the World Wide Web, your computer tells another computer which tells another computer that you want such and such document delivered to your computer, and it arrives in seconds. Some geographers measure and map cyberspace by looking at where most of the Internet's traffic flows though, to, and from.

What is **geographic illiteracy**?

In 1989, the National Geographic Society commissioned a survey to find out how much Americans and residents of several other countries knew about the world around them. Unfortunately, American youths scored the worst. Swedes knew the most about world geography. The media subsequently reported the "geographic illiteracy" of the American population. Due to the attention given to this problem, geographic education has since become a greater priority for educators.

What is the **AAG**?

The Association of American Geographers (AAG), is a professional organization of academic geographers and geography students. The AAG was founded in 1904 and publishes two key academic journals in geography, the *Annals of the Association of American Geographers* and the *Professional Geographer.* The AAG also holds annual conferences and supports regional and specialty groups of geographers.

What is the **NCGE**?

The National Council for Geographic Education (NCGE) is an organization of educators that seeks to promote geographic education. The NCGE publishes the *Journal of Geography* and holds conferences every year.

What is the **National Geographic Society**?

Founded in 1888, the National Geographic Society has supported exploration, cartography, and discovery and publishes the popular magazine *National Geographic*, the fifth most-popular magazine in the United States.

What is a good way to **learn where places are**?

The best way to learn where places are in the world is to use an atlas like a dictionary. Any time you hear of an unfamiliar place on television or while reading, look it up in an atlas. Before long, you'll be extremely geographically literate.

What do **modern geographers** do?

While there are a few jobs with the title of "geographer," many geography students use their analytical ability and knowledge of the world to work in a variety of fields. Geography students often take jobs in fields such as city planning, cartography, marketing, real estate, environment, and teaching.

Performed by Iohn Speed, and are to be sold by
Thomas Baßett in Fleetstreet, and Richard
Chiswell in S.t Pauls Churchyard.

What makes a piece of paper a **map**?

No matter what the medium, all maps must be a representation of an area of the Earth, celestial bodies, or space. Though maps are commonly printed on paper, they can come in a variety of forms, from being drawn in the sand to being viewed on computers. A map should have a legend (a guide explaining the map's symbols), a notation of which way is north, and an indicator of scale. No map is perfect and every map is unique.

How do **cartographers** shape our world?

Cartographers are map makers and cartography is the art of map making. Cartographers map neigborhoods, cities, states, countries, the world, and even other planets. There are as many types of maps to make as there are cartographers to make them.

Who would purposely create **false maps**?

Within the Soviet Union, incorrect maps were produced as a matter of course. Soviet maps purposely showed the locations of towns, rivers, and roads in incorrect places. Often, in different editions of the same map, towns would disappear from one version to the next. Street maps of Moscow were particularly incorrect and non-proportional. The cartographic deceit of the U.S.S.R. was an effort to keep the geography of the

A cartographer at work. (Roger Ressmeyer/Corbis)

country a secret, not only from foreigners but also from its own citizens. Within the U.S.S.R., even official government agencies were not allowed to have accurate maps.

Who decides which **names** go **on maps**?

In the United States, the U.S. Board on Geographic Names (BGN) approves the official names and spellings of cities, rivers, lakes, and even foreign countries. If a town would like to change its name, it must petition the BGN for approval. Upon approval, the name is officially changed and updated in federal government gazetteers and records, which official and commercial mapmakers use for their maps.

Why is a book of maps called an **atlas**?

The term "atlas" comes from the name of a mythological Greek figure, Atlas. As punishment for fighting with the Titans against the gods, Atlas was forced to hold up the planet Earth and the heavens on his shoulders. Because Atlas was often pictured on ancient books of maps, these became known as atlases.

How can **maps** be used to **start wars**?

Prior to their 1990 invasion of Kuwait, Iraq produced official maps that showed the independent country of Kuwait as Iraq's 19th province. Iraq used these maps as justification for their 1990 invasion and attempted annexation of Kuwait (Iraq was actually after Kuwait's oil reserves). Maps have been, and still are, used by a multitude of countries, provinces, and cities to prove ownership of a certain piece of land.

How does a **sextant** help navigators?

In 1730, the sextant was invented independently by two men, John Hadley and Thomas Godfrey. Using a telescope, two mirrors, the horizon, and the sun (or another celestial body), the sextant measures the

How do I get to the refrigerator in the dark?

Not all maps are written on paper. When trying to reach the refrigerator at night, we do not smell our way to food, we use a map based on our memory of the room. If we stumble on our way, it is usually over a misplaced toy or shoe that we did not remember leaving there. Everyone has these kinds of maps in his or her mind. These mental maps help you find your way not only to the refrigerator in the dark, but also to the grocery store and to work. People not only have mental maps of common trips they make, but also of their city, country, and even the world. Every person's mental map is unique, based on how wide an area that person travels and their knowledge of the world.

angle between the horizon and the celestial body. With this measurement, navigators could determine their latitude while at sea.

When was the **compass** invented?

As early as the 11th century, the Chinese were using a magnetic needle to determine direction. At approximately the same time, the Vikings may have also used a similar device. A compass is simply a magnetic needle that points toward the magnetic north pole.

How old is the **oldest known map**?

Circa 2,700 B.C.E., the Sumerians drew sketch maps in clay tablets that represented their cities. These maps are the oldest known maps.

What is the oldest known **map drawn to scale**?

In the sixth century B.C.E, the Greek geographer Anaximander created the first known map drawn to scale. His map was circular, included known parts of Europe and Asia, and placed Greece at its center.

Where is the **equator**?

The equator is the line located equidistant between the North and South Poles. The equator evenly divides the Earth into the northern and southern hemispheres and is zero degrees latitude.

LATITUDE AND LONGITUDE

What are **latitude and longitude** lines?

Lines of latitude and longitude make up a grid system that was developed to help determine the location of points on the Earth. These lines run both north and south and east and west across the planet. Lines of latitude (those that run east and west) begin at the equator, which is zero degrees. They extend to the North Pole and the South Pole, which are 90 degrees north and 90 degrees south, respectively. Lines of longitude (those that run north and south) begin at the Prime Meridian, which is the imaginary line that runs through the Royal Observatory in Greenwich, England. The lines of longitude extend both east and west from the Prime Meridian, which is 0 degrees, and converge on the opposite side of the Earth at 180 degrees.

Are lines of longitude and latitude all the **same length**?

No, they are not. Only the lines of longitude are of equal length. Each line of longitude equals half of the circumference of the Earth because each extends from the North Pole to the South Pole. The lines of latitude are not all equal in length. Since they are each complete circles that remain equidistant from each other, the lines of latitude vary in size from the longest at the equator to the smallest, which are just single points, at the North and South Poles.

How wide is a degree of **longitude**?

Though there are only a couple dozen lines of longitude shown on most globes and world maps, the Earth is actually divided into 360 lines of longitude. The distance between each line of longitude is called a degree. Because the lines of longitude are widest at the equator and converge at the Poles, the width of a degree varies from 69 miles wide to zero, respectively.

How **wide** is a degree of **latitude**?

Though there are only about a dozen lines of latitude shown on most globes and world maps, the Earth is actually divided into 180 lines of latitude. The distance between each line of latitude is called a degree. Each degree is an equal distance apart, at 69 miles.

What do **minutes and seconds** have to do with longitude and latitude?

Each degree of longitude and latitude is divided into 60 minutes. Each minute is divided into 60 seconds. An absolute location is written using degrees (°), minutes ('), and seconds (") of both longitude and latitude. Thus, the Statue of Liberty is located at 40° 41' 22" north, 74° 2' 40" west.

Which comes first, latitude or longitude?

Latitude is written before longitude. Latitude is written with a number, followed by either "north" or "south" depending on whether it is located

The Prime Meridian at Greenwich, England. (Dennis di Cicco/Corbis)

north or south of the equator. Longitude is written with a number, followed by either "east" or "west" depending on whether it is located east or west of the Prime Meridian.

How can I **remember** which way latitude and longitude run?

You can remember that the lines of latitude run east and west by thinking of lines of latitude as rungs on a ladder ("ladder-tude"). Lines of longitude are quite "long" because they run from the North Pole to the South Pole.

How can a **gazetteer** help me find latitude and longitude?

A gazetteer is an index that lists the latitude and longitude of places within a specific region or across the entire world. Many atlases include a gazetteer, and some are published separately.

How can I find the latitude and longitude of a **particular place**?

To find latitude and longitude of a particular location, you will need to consult either a gazetteer or a computer database that includes longitude and latitude data. Though gazetteers are readily accessible, they don't include as many places as online databases. There are a number of sites on the Internet that have extensive databases of latitude and longitude and even include such specific places as public buildings.

Why was the Prime Meridian established at **Greenwich**?

In 1675, the Royal Observatory in Greenwich, England, was established to study determination of longitude. In 1884, an international conference established the Prime Meridian as the longitudinal line that passes through the Royal Observatory. The United Kingdom and United States had been using Greenwich as the Prime Meridian for several decades

before the conference.

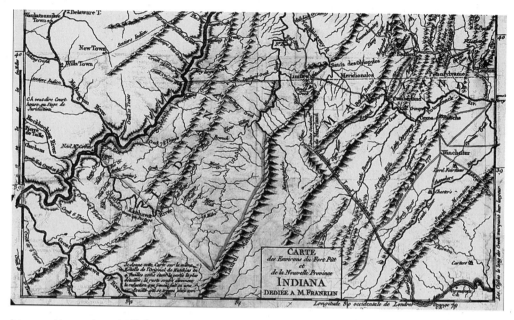

A topographic map shows detailed man-made and physical features. (Library of Congress/Corbis)

READING AND USING MAPS

What is the difference between a **physical and a political** map?

A physical map shows natural features of the land such as mountains, rivers, lakes, streams, and deserts. A political map shows human-made features and boundaries such as cities, highways, and countries. The maps we use in atlases and see on the walls of classrooms are typically a combination of the two.

What is a **topographic** map?

A topographic map shows human and physical features of the Earth and can be distinguished from other maps by its great detail and by its contour lines indicating elevation. Topographic maps are excellent sources of detailed information about a very small area of the Earth. The United States Geological Survey (USGS) produces a set of topographic maps for the United States that are at a scale of 1: 24,000 (one inch equaling

27

2,000 feet). You can purchase these maps online, at sporting goods stores, or through the USGS itself.

Why are road maps so **difficult to fold**?

The problem lies with the multitude of folds required to return the map to its original, folded shape. The easiest way to fold a road map is to study the creases and to fold the map in the order that the creases will allow. But once you've made a mistake, the folds have lost their tell-tale instructions. To fold a road map, begin by folding it accordion style, making sure that the "front" and "back" of the folded design appear on top. Then, once the entire map is folded accordion style, fold the remaining slim, long, folded paper into three sections. And, violà, your road map is folded!

Why is **color** important on a relief map?

A relief map portrays various elevations in different colors. But, a common color scheme found on relief maps causes a problem. On these maps, mountains are displayed as red or brown while lowlands are shown in shades of green. This is confusing because the green areas on the map are often misconstrued as fertile land while brown areas are mistaken for deserts. For example, an area such as California's Death Valley, which is shown in green on relief maps because it lies below sea level, seems fertile, when actually it is an inhospitable desert.

What does the **scale of a map** tell me?

A scale indicates the level of detail and defines the distances between objects on a map. On a map, scales can be written as a fraction, a verbal description, or as a bar scale.

A fraction, or ratio, using the example of 1/100,000 or 1:100,000, indicates that one unit of any form of measurement on the map is equivalent to 100,000 units of the same measurement in the area being represented. For instance, if you use inches as the unit of measurement, then one inch on the map would equal 100,000 inches in the area represented by the map.

A verbal description describes the relationship as if it were a verbal instruction, such as "one inch equals one mile." This allows the versatility of having different units of measurement.

A bar scale uses a graphic to show the relationship between distance on the map to distance in the area represented. The bar scale is the only type of scale that allows a reduction or enlargement of the map without distorting the scale. This is because when you increase the size of the map, the bar scale is increased proportionally. For a fraction or verbal scale, the proportion (1:1000) is only true for the map at that size. For example, when enlarging a map, the map might become twice as large but the numbers in a ratio do not change, as they would need to in order to stay accurate.

How can I determine the **distance between two places** by using a scale?

By using a ruler, compare the distance between two points on a map with the information on the scale to calculate the actual distances between the two points. For example, if you measure the distance between two towns as being five inches and the ratio says 1:100,000, then the actual distance between the towns is 500,000 inches (7.9 miles).

What is the difference between **small and large** scale maps?

A small-scale map shows a small amount of detail over a wide area, such as the world. A large-scale map shows a large amount of detail while representing a limited area, such as neighborhoods or towns.

Why is every map **distorted**?

No map is completely accurate because it is impossible to accurately represent the curved surface of the Earth on a flat piece of paper. A map of a small area usually has less distortion because there is only a slight curve of the Earth to contend with. A map of a large area, such as maps

How is the Earth's surface like an orange peel?

All attempts to represent a sphere, like the Earth, in a flat representation result in distortions. The Earth's surface is like an orange peel. If one were able to peel an orange in one piece and then try to flatten the peel, cracks and tears would appear. Attempting to "peel" the Earth and then lay that information on a flat surface in a map creates these same open areas. Map makers attempt to create maps that represent the spherical Earth with as little distortion as possible. The various strategies for this are called projections.

of continents or the world, are significantly distorted because the curvature of the Earth over such a large area is extreme.

Why does **Greenland appear larger** on most maps than it actually is?

Because of the distortions that must appear on all maps, many maps place the distortion in the northern and southern extremes of the Earth. In one of the common projections, known as Mercator, Greenland appears to be similar in size to South America, despite the fact that South America is actually eight times larger than Greenland. The advantage of the Mercator projection is that the lines of latitude and longitude remain perpendicular, thus the map is useful for navigation.

How can a **legend** help me read a map?

The legend, usually found in a box on the map, is information that explains the symbols used on a map. Though some symbols seem standard, like a railroad line, even those can be represented differently on different maps. Since there really are no standard symbols, each maps' legend should be consulted when reading a map.

What is a **compass rose**?

On old maps, the directions of the compass were represented by an elaborate symbol, known as a compass rose. Many of the older compass roses displayed 32 points, representing not only the four cardinal directions (north, south, east, and west) but also 28 subdivisions of the circle (south-west, south south-west, etc.). This directional symbol resembled a rose, hence its name. Though compasses are now often drawn with only the four cardinal directions and the resemblance to the flower is minimal, the directional symbol is still called a compass rose.

A compass rose. (Corbis-Bettmann)

Why is there often a **cross** next to the east direction on maps?

On old maps, a cross often sits next to the east direction on a compass rose. This cross represents the direction to Paradise and the Holy Land.

Where can I **buy maps**?

There are many places you can buy maps. Most large bookstores offer an extensive collection of local and foreign travel maps, wall maps, and atlases. Also, many cities have specialty travel and map stores that offer a larger and more varied collection of maps, as well as maps of more exotic locales.

But what if I **can't find** the map I'm looking for?

Not all maps can be found at bookstores or even in specialty stores. If you are looking for an extremely specific and relatively uncommon map,

31

visit a local university's map collection. Their collections are often far greater in size and breadth than any store. If you need help locating a map, you should be able to discuss your map needs with a friendly map librarian.

What is the difference between **relative and absolute** location?

There are two different ways to describe where a place is located; relative location and absolute location. Relative location is a description of location using the relation of one place to another. For instance, using relative location to describe where the local video store is you might say that it's on Main Street, just past the high school. Absolute location describes the location of a place by using grid coordinates, most commonly latitude and longitude. For instance, the local video store would be described as being located at 23 degrees, 23 minutes, 57 seconds N and 118 degrees, 55 minutes, 2 seconds W.

MODERN MAPPING

What are **satellites** photographing?

Satellites capture images of the Earth's weather patterns, the growth of cities, the health of plants, and even individual buildings and roads. Satellites circle the Earth, or remain geostationary (in the same place with respect to the Earth), and send data back to the Earth via radio signals.

How have satellites **changed map making**?

Satellite images, which are accurate photographs of the Earth's surface, allow cartographers to precisely determine the location of roads, cities, rivers, and other features on the Earth. These images help cartographers create maps that are more accurate than ever before. Since the Earth is a dynamic and ever-changing place, satellite images are great tools that allow cartographers to stay up-to-date.

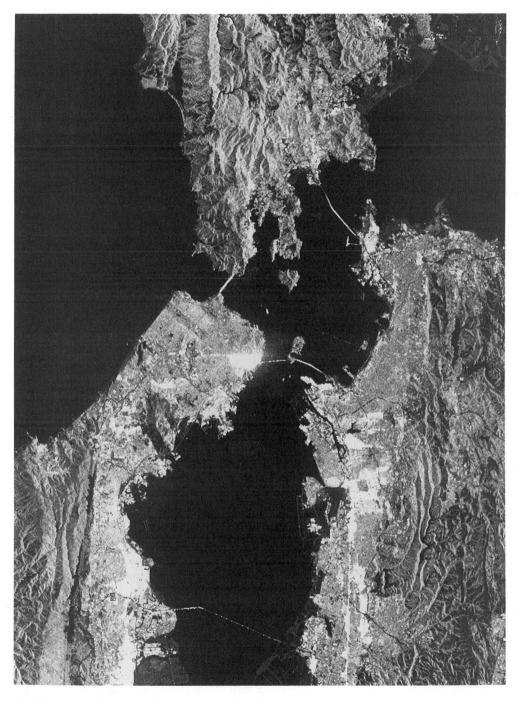

A satellite image of San Francisco Bay. (NASA-JPL/Corbis)

How much **junk** is there in space?

In addition to operational satellites, there are approximately 8,800 pieces of space junk surrounding the Earth, from tiny screws to booster rockets. There are plans for the future to build a radar system that would track every piece of space junk so that space vehicles and satellites can avoid irreparable damage.

How has **GIS** revolutionized cartography?

Geographic Information Systems (GIS) began in the 1960s with the popularity of computers. Though very simplistic in its beginning, new technology and inventions have expanded and enhanced the functions of GIS. GIS has revolutionized cartography by using computers to store, analyze, and retrieve geographic data, thus allowing infinite numbers of comparisons to be made quickly. The program formulates information into various "layers," such as the location of utility lines, sewers, property boundaries, and streets. These layers can be placed together in a multitude of combinations to create a plethora of maps, unique and suitable to each individual query. The versatility of GIS makes it indispensable to local governments and public agencies.

How can GIS **help my town**?

Your community can use GIS on a day-to-day basis and in emergency situations. GIS allows public works departments, planning offices, and parks departments to monitor the status of the community's utilities, roads, and properties. In an emergency, GIS can give emergency teams the information they need to evacuate endangered areas and respond to the crisis.

How does a **GPS** unit know where I am?

Individual Global Positioning System (GPS) units on the Earth receive information from a U.S. military-run system of 24 satellites that circle the Earth and provide precise time and location data. The individual GPS unit receives data from three or more satellites that triangulate its

How did a map stop cholera?

In the 1850s an outbreak of cholera threatened London. Dr. John Snow, a British physician, mapped the deaths associated with the disease and determined that many deaths were occurring near one water pump. The pump handle was removed and the spread of the disease stopped. Prior to this time, the method by which cholera spread was unknown. Today, medical geographers and epidemiologists frequently use cartography to determine the cause and spread of disease or epidemics.

absolute location on the Earth's surface. If you are carrying such a device, your absolute location is the same as that of the device.

How can GPS keep me from getting lost?

A GPS unit provides precise latitude and longitude for the location of the device. By using a hand-held GPS unit along with a map that provides latitude and longitude (such as a topographic map), you can determine your precise location on the Earth's surface. This is a valuable tool for those who hike or travel in remote regions and for ships at sea.

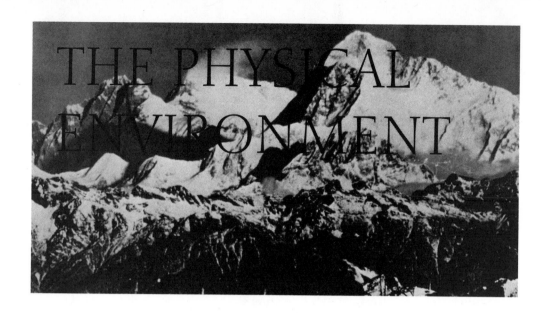

THE PHYSICAL ENVIRONMENT

GEOLOGIC TIME

What is **geologic time**?

Geologic time is a time scale that divides the history of the planet Earth into eras, periods, and epochs from the birth of the planet to the present. The oldest era is the Precambrian, which began 4.6 billion years ago and ended about 570 million years ago. Next came the Paleozoic Era, which lasted from 570 to 245 million years ago, followed by the Mesozoic Era, from 245 to 66 million years ago. We're now living in the Cenezoic Era, which began 66 million years ago. The Paleozoic, Mesozoic, and Cenozoic eras are each divided into periods. Additionally, the Cenezoic Era is divided into even smaller units of time called epochs. The last ten thousand years (the time since the last significant Ice Age) is called the Holocene Epoch.

How **old** is the Earth?

The Earth is approximately 4.6 billion years old.

How was the **Earth formed**?

Scientists believe that the Earth was formed, along with the rest of the solar system, from a massive gas cloud. As the cloud solidified, it formed the solid masses such as the Earth and the other planets.

Did an asteroid kill the **dinosaurs**?

Approximately 65 million years ago, a six-mile-wide asteroid struck the Earth. This impact might have started a chain of events that led to the extinction of about two-thirds of the Earth's species of animals and plants, including the dinosaurs. An asteroid that large would have created a layer of dust that would have surrounded the Earth, lowering temperatures, and causing deadly, highly acidic rain.

Will another large asteroid **strike the Earth**?

Yes, asteroids have struck the Earth in the past and are likely to strike again in the future. Small asteroids strike the planet about every 1,000 to 200,000 years. Large asteroids are much less likely to threaten the planet; they strike only about three times every million years. Huge asteroids, such as the one that may have killed dinosaur species, are even less frequent.

How many places on Earth have **evidence of asteroid impacts**?

There are about 150 places on Earth that display evidence of asteroid impacts. Depending on the size of an asteroid, it can have significant effects on the planet. If an asteroid strikes the ocean, it can create huge destructive tsunamis or tidal waves. If an asteroid strikes land, it can create a huge crater, cause earthquakes, and propel debris into the atmosphere, creating major climatic changes.

THE EARTH'S MATERIALS AND INTERNAL PROCESSES

How thick is the **Earth's crust**?

The thickness of the Earth's crust varies at different points around the planet. Under continents, the crust is approximately 15 miles thick, but under the oceans it is a mere 5 miles thick.

A 50,000 year old asteroid crater in Arizona. (Jonathan Blair/Corbis)

What is **continental drift**?

The Earth is divided into massive pieces of crust that are called tectonic plates. These plates lie wedged together like a puzzle. The plates slowly move, crashing into each other to form mountain ranges, volcanoes, and earthquakes. The plates are like rafts floating on water; this is called continental drift.

How many **tectonic plates** are there?

There are a dozen significant plates on the planet. Some of the largest include the Eurasian Plate, North American Plate, South American Plate, African Plate, Indo-Australian Plate, Pacific Plate, and Antarctic Plate. Some smaller plates are located between the major plates. The smaller plates include the Arabian Plate (containing the Arabian Peninsula), the Nazca Plate (located to the west of South America), the Philippine Plate (located southeast of Japan, containing the northern Philippine islands), the Cocos Plate (located southwest of Central America), and the Juan de Fuca Plate (just off the coast of Oregon, Washington, and Northern California).

What was Pangea?

Pangea, which existed about 250 million years ago, was one huge continent, including the land of all seven continents. It was located near present-day Antarctica and has slowly drifted and split to form the continents as we know them today. The continents and their tectonic plates continue to move and will one day be in a much different arrangement than they are today.

How are **mountains** formed?

The process of orogeny, or mountain building, is related to continental drift. When two tectonic plates collide, they often form mountains. The Himalayas are the result of the Indo-Australian plate colliding with the Eurasian plate. At these collision zones, volcanoes and earthquakes are common.

How did the **Himalayas** form?

About 30 to 50 million years ago, the landmass of India pressed into the landmass of Asia, pushing up land at the place of impact and creating the Himalayas. Even today, as the Indian subcontinent presses against Asia, the Himalayas continue to grow and change.

What type of rocks are **formed by lava**?

Igneous rocks are formed when liquid magma under the surface of the Earth, or lava on the surface of the Earth, cools and hardens into rock.

What type of rocks are **formed from particles**?

Sedimentary rocks are formed by the accumulation and squeezing together of layers of sediment (particles of rock or remains of plant and

The Himalayas as seen from space.
(NASA/Corbis)

animal life) at the bottom of rivers, lakes, and oceans or even on land. The continual accumulation of more and more layers of sediment places a great amount of pressure on the lowest layers of sediment and, over time, compresses them into rock.

What are **recycled rocks**?

Metamorphic rocks are recycled rocks. Metamorphic rocks are rocks that had a prior existence as sedimentary, igneous, or even another

41

Oil, like that being mined here, is a nonrenewable resource that took millions of years to create. (Archive Photos)

metamorphic rock. Underground heat and pressure metamorphose one type of rock into another, creating a metamorphic, or reclycled rock.

Which is **larger, clay or sand**?

A single grain of sand is 1500 times larger than a grain of clay.

What is a **dike**?

A dike is magma that has risen up through a crack between layers of rock. When this magma solidifies, it becomes very solid rock. If the rock around it is eroded, a dike can form great rock monoliths above the ground.

What are **hot springs**?

Hot springs are created by underground water that is heated and percolates to the Earth's surface. Aside from being natural baths, the steam from hot springs can be used to drive turbines, which create electricity. This type of energy production is called geothermal energy.

Why does the **ground sink**?

In many places over the world, seemingly solid land lies over vast oil deposits or water aquifers. Without the liquid supporting it, the ground sinks into the space left behind. In some parts of California's Imperial Valley, the land has dropped more than 25 feet due to underground water being removed from the area. Unless the pumping of underground water and oil is stopped, the land will continue to sink.

Why do houses fall into **sinkholes**?

Houses that sit upon limestone rock have the proclivity to fall into sinkholes. As underground water wears away the limestone rock, it creates underground caverns. If the water wears away too much limestone, the cavern may collapse, taking anything on the surface with it. A sinkhole is just one of the many reasons to have your home inspected by a geologist.

NATURAL RESOURCES

What is a **renewable resource**?

A renewable resource is one that can be replenished within a generation. Forests, as long as they are replanted, are a renewable resource. Materials such as oil, coal, and natural gas are known as nonrenewable resources because they require millions of years to be created. So, once the world's supply of oil is gone, it's gone for a long, long time.

What are **fossil fuels**?

Underground fuels such as natural gas, oil, and coal are all known as fossil fuels because they are encased in rocks, just like fossils. It takes millions of years of compression and the building up of dead plants and animals to create these fuels.

What is a **fossil**?

The outline of the remains of a plant or animal embedded in rock is called a fossil. Fossils are formed when a plant or animal dies and becomes covered up by sediments. Over time, the layers compress the remains, which are then embedded into the rock.

LANDSCAPES AND ECOSYSTEMS

What are **basins and ranges**?

Basins and ranges are sets of valleys and mountains that are spaced close together. Most of Nevada and western Utah is composed of sets of basins and ranges.

What is **permafrost**?

Soil that is permanently (or for a good part of the year) frozen is known as permafrost. Permafrost occurs in higher latitudinal regions, which have cold climates.

How much of the Earth's surface is **frozen**?

About one-fifth of the planet is permafrost, or frozen for all or most of the year.

What is a **jungle**?

A jungle is a forest that is composed of very dense vegetation. The term tropical rain forest can often be used interchangeably with jungle. Jungles occur most often in tropical areas such as the Amazon and Congo river basins.

How can an area be lower than sea level?

Land in the midst of a continent can be lower than sea level because the land is not close enough to the sea to be flooded with water. Movement of the tectonic plates pushes areas like the Dead Sea in Israel and Death Valley in California to elevations lower than sea level.

What is a **rain forest**?

A rain forest is any densely vegetated area that receives over 40 inches of rain a year.

What is a **tropical rain forest**?

A tropical rain forest is a rain forest that lies between the Tropic of Cancer in the north and the Tropic of Capricorn in the south (i.e. within the "tropics"). Tropical rain forests are known for their very diverse species of plant and animal life. Tropical rain forests exist throughout Central America, northern Brazil, the Congo River basin, and Indonesia.

What is a **desert**?

A desert is an area of light rainfall. Deserts usually have little plant or animal life due to the dry conditions. Contrary to the popular image, deserts are not just warm, sand-swept areas like the Sahara Desert, but can also be frigid areas like Antarctica, one of the driest places on Earth.

What was the **highest temperature** ever recorded?

The world's highest temperature was recorded in the desert in Libya in 1922. The temperature was 136 degrees Fahrenheit. California's Death

Valley holds the record for the highest temperature in the United States, 134 degrees in 1913.

Do **oases** really exist?

Oases do exist and they are quite prevalent throughout the eastern Sahara Desert in Africa. An oasis has a source of water, often a spring, that allows vegetation to grow. Small towns are located at some larger oases in the desert. Oases have been traditional stopping places for nomads travelling across deserts.

How do **sand dunes move**?

Sand dunes are created and transported by wind. Wind blows sand from the windward side of the dune to the opposite side, slowly transporting it across the landscape.

What causes **erosion**?

Wind, ice, and water are the most common agents of erosion. They wear down and carry away pieces of rock and soil. The process is accelerated when trees that help hold the soil in place have been destroyed by fire or have been chopped down. With fewer trees, the soil is easily eroded and washed away, leaving a barren surface where plants can no longer grow.

What does a **glacier leave behind**?

When a glacier moves across the land, it acts like a giant bulldozer, pushing and collecting rock, dirt, and debris. A moraine is a deposit of rock and dirt carried by a glacier and left behind once the glacier melts and recedes.

What is a **tree line**?

A tree line is the point of elevation at which trees can no longer grow. The tree line is caused by low temperatures and frozen ground (permafrost).

A fjord in Norway. (Archive Photos)

How do **forest fires** help forests?

An occasional fire is often necessary for a forest. Forest fires clear undergrowth, giving more room for trees to grow, thus rejuvenating the forest. Since forest fires are usually extinguished by fire fighters as rapidly as possible, the amount of undergrowth in forests has increased. This extra undergrowth can become extremely flammable, making fires even more dangerous to people. A policy of allowing the forest to burn naturally, while protecting human structures, produces a more natural environment.

What is **tundra**?

Tundra is dry, barren plains that have significant areas of frozen soil or permafrost. Tundra is common in the northernmost parts of North America, Greenland, Europe, and Asia. Although rather inhospitable, there is plant life on the tundra. This life consists of low, dense plants such as shrubs, herbs, and grasses. There are even some species of insects and birds that can survive the harsh conditions of tundra.

What is a **fjord**?

During the ice ages, glaciers, which were prevalent at higher latitudes and elevations, became so large that gravity drove them to lower elevations, eventually all the way to the sea. On their way, glaciers would carve deep canyons in the surface of the Earth. At the end of the ice age, as the ice melted and the ocean level rose, these glacial troughs filled with seawater. These very dramatic-looking canyons with high cliffs hanging over a thin bay of water are known as fjords. Fjords are very common in Norway and Alaska.

WATER AND ICE

How much of the **Earth is covered by water**?

About 70 percent of the surface area of the Earth is covered by water. The other 30 percent of the Earth is land located primarily in the Northern Hemisphere. If you look at a globe, you'll notice that the Southern Hemisphere has a great deal of ocean.

How does the **hydrologic cycle** work?

The movement of water from the atmosphere to the land, rivers, oceans, and plants and then back into the atmosphere is known as the hydrologic cycle. We can pick an arbitrary point in the cycle to begin our examination. Water in the atmosphere forms clouds or fog and falls (precipitates) to the ground. Water then flows into the ground to nourish plants, or into streams that lead to rivers and then to oceans, or it can flow into the groundwater (underground sources of water). Over time, water sitting in puddles, rivers, and oceans is evaporated into the atmosphere. Water in plants is transpired into the atmosphere. The process of water moving into the atmosphere is collectively known as evapotranspiration.

What is **evapotranspiration**?

Evapotranspiration is the combination of water vapor being evaporated from the surface of the Earth (such as from lakes, rivers, or puddles)

Where is all the water?

Over 97 percent of the world's water lies in the oceans and is too salty to drink or to irrigate crops with (except when the water is cleaned through a desalinization plant, which is not done very often). About 2.8 percent of the world's water supply is fresh water. Of that 2.8 percent, about 2 percent is frozen in glaciers and ice sheets. This leaves only about 0.8 percent of the world's water that is accessible through aquifers, streams, lakes, and in the atmosphere. The water that we use primarily comes from this 0.8 percent.

into the atmosphere, and transpiration, which is the movement of water from plants to the air.

What is an **aquifer**?

An aquifer is an underground collection of water that is surround by rock. The creation and filling of an aquifer is a very slow process, as it relies upon water to percolate through the soil and rock layers and into the aquifer. An aquifer lies above a lower layer of rock that holds the water in place and keeps it from moving further underground.

What is the **Oglala Aquifer**?

The Oglala Aquifer is a huge aquifer that spans an area from Texas to Colorado and Nebraska. The oldest water deposited in the aquifer is over one million years old, and only a very small amount water is added each year. The Oglala Aquifer is being pumped rapidly by the farms in the region, causing a reduction in the amount of water in the aquifer. Consequently, wells have to be continually deepened so that they can continue to pump water.

Will there be another ice age?

Yes, eventually the Earth will again cool and ice will cover land at higher latitudes and elevations. It may be a hundred years from now or it may be thousands of years away, but the Earth's climate is always slowly changing.

Why are we losing **ground water**?

Water is pumped from aquifers around the world for irrigation, industrial, and household needs. Aquifers do not refill as rapidly as water is being pumped out, so in many areas there is a danger that some aquifers may disappear altogether.

What are **ice ages**?

Throughout the life of the planet, the climate has warmed and cooled many times. During the cooling periods, ice ages have occurred. During the ice ages, large sheets of ice cover large portions of land. In the most recent ice age, which ended about 10,000 years ago, large parts of northern Europe and North America were covered by ice sheets.

What is the **Coriolis** effect?

Due to the rotation of the Earth, any object on or near the Earth's surface will veer to the right in the Northern Hemisphere and to the left in the Southern Hemisphere. This applies especially to phenomena such as ocean currents and wind. Imagine a missile being fired at New York by Los Angeles. As the missile flies over the United States, the Earth continues to rotate under the missile and it strikes New Jersey instead. Missile launchers and pilots need to factor the spinning of the Earth into their trajectories in order to end up in the right place. North of the equator, ocean currents and winds rotate clockwise, but south of the equator, the opposite is true.

Does the Coriolis effect make the water in my **toilet, sink, and bathtub swirl clockwise**?

No, the Coriolis effect has very little effect on such small bodies of water. The flow down the drain is mostly a function of the shape of the container.

If I keep walking in a straight line, will the **Coriolis effect cause me to veer**?

If your body were completely symmetrical (and no one's is) and neither leg were longer and you were walking on perfectly flat land then yes, you might start veering due to the Coriolis effect.

What is the difference between **a bay and a gulf**?

Both are bodies of water partially surrounded by land, and a bay is a smaller version of a gulf. Famous bays include the San Francisco Bay (California), the Bay of Pigs (Cuba), Chesapeake Bay (Maryland/Virginia area), Hudson Bay (Canada), the Bay of Bengal (a large bay near India and southeast Asia), and the Bay of Biscay (France). Famous gulfs include the Gulf of Mexico (southern U.S.), the Persian Gulf (between Saudi Arabia and Iran), and the Gulf of Aden (between the Red Sea and the Arabian Sea).

Where does the **Loch Ness Monster** live?

The fabled monster is supposed to live in Loch Ness. The term "loch" is Gaelic and is used in Scotland to refer to a lake or narrow inlet of the sea. Loch Ness is fully surrounded by land and is therefore a lake.

How are **waves** created?

Waves are created by wind blowing across the surface of the water. Though waves appear to move along the surface of the water, they are simply the movement (oscillation) of water up and down due to the fric-

Old Faithful geyser in Yellowstone National Park, Wyoming. (Archive Photos)

Niagara Falls. (Library of Congress/Corbis)

tion of the air. When waves occur near the shore, they may become steeper and "break."

How does **Old Faithful** shoot water into the air?

A geyser, such as Yellowstone National Park's famous Old Faithful, is the result of an underground aquifer that is warmed by heated rocks and magma underground. There is a small fissure or crack in the aquifer's surface that allows the steam and heated water to jet from the ground (about every hour at Old Faithful).

How does water **wash away the land**?

Drops of rain hit soil and rock and displace grains. When water flows over the surface, it loosens and carries away pieces of rock or soil. There is a tremendous amount of energy in a raindrop. Over days, weeks, months, years, centuries, and millennia, the erosive power of water can cut through even the strongest rocks. The material that the flowing

by sending warm water from the Caribbean northeast across the Atlantic Ocean to northern Europe. A current known as the Antarctic Circumpolar Current circles the southern continent. The North Atlantic and North Pacific oceans each have a large clockwise current, while the South Atlantic and South Pacific oceans each have a large counterclockwise current.

What are the **largest seas**?

Parts of oceans that are surrounded by islands or otherwise partially enclosed are often known as seas. The five largest seas in order are: the South China Sea, the Caribbean Sea, the Mediterranean, the Bering Sea, and the Gulf of Mexico.

Why is the Mediterranean Sea **so salty**?

Due to the high temperatures in the Mediterranean region, evaporation of the Mediterranean Sea occurs more rapidly than in other bodies of water, therefore more salt is left behind. The warm, dense, salty water in the Mediterranean is replaced by less salty and dense Atlantic water in the Strait of Gibraltar. Water that flows into the Mediterranean from the Atlantic Ocean usually remains in the Sea for anywhere from 80 to 100 years before returning to the Atlantic Ocean.

Has the **Mediterranean Sea** always been there?

Salt and sediment found at the bottom of the Mediterranean Sea prove that on several occasions the Mediterranean Sea has dried up, leaving a large layer of salt behind. Scientists speculate that the Strait of Gibraltar has, on occasion, closed up, keeping water from being able to flow back and forth between the Atlantic Ocean and the Mediterranean Sea.

What are the **seven seas**?

The "seven seas" spoken of by mariners from long ago are oceans or parts of oceans. The Atlantic and Pacific Oceans are so large that they were each divided into two "seas." The Antarctic, Indian, and Arctic were also considered seas, thus totaling seven. If a sailor had sailed upon all seven seas, he had sailed around the world. There are not just seven but dozens of seas in the world.

Where are the four **colored seas**: Black, Yellow, Red, and White?

The four colored seas are not geographically associated with one another. The Black Sea is located near the Balkan Peninsula and is bordered by Turkey, Russia, and Ukraine (it is also the home of the port city of Odessa). The Red Sea is located to the south of the Black Sea, between the Arabian Peninsula (Saudi Arabia), and Africa. The Red Sea has been a major trade route for hundreds of years and has been especially useful since the completion of the Suez Canal. The White Sea is in northern Europe. It is part of the Arctic Ocean and is a Russian sea (it lies to the east of Finland). The Yellow Sea is far to the east, between China and the Korean Peninsula.

Is the Black Sea **really black**?

No, it is not. This sea, located to the north of Turkey, is quite deep and has darker looking water than most water bodies, but receives its name from the inhospitableness of the waters for sailing.

What is a **desalinization plant**?

A desalinization plant pumps ordinary seawater through a variety of expensive processes, transforming the salty water into fresh water. This process has been used with some success in Texas, the Caribbean, and the Middle East. It is much more efficient and less expensive, however, to clean waste water (water that has been used for bathing, cooking, cleaning, etc.) than it is to clean and desalinate seaweater.

RIVERS AND LAKES

What is the **longest river** in the world?

Egypt's famous Nile River is the longest in the world. It is more than 4,100 miles long from its sources in the Ethiopian Highlands (the source of the portion of the Nile called the Blue Nile) and Lake Victoria (the source of the White Nile). The Nile Valley is the center of contemporary and ancient Egyptian civilization. Following the Nile in length are the Amazon (in Brazil), the Missouri-Mississippi (U.S.), the Chang or Yangtze (China), the Huang or Yellow (China), and the Ob (Russia).

What is the **longest river in the U.S.**?

The Missouri-Mississippi River is the longest in the U.S., approximately 3860 miles in length.

Why are the **Missouri-Mississippi Rivers** lumped together?

Actually, the Missouri River was incorrectly named. The Missouri River is actually the main feeder river of what is now known as the Mississippi River. Usually, the main feeder bears the same name as the rest of the river. Therefore, the full length of the Mississippi River, including the Missouri River, is known as the Missouri-Mississippi River.

Which river carries the **most water**?

By far, Brazil's Amazon River carries more water to the sea than any other river in the world. The discharge at the mouth of the river is about seven million cubic feet per second, which is about four times the flow of the Congo in Africa, the river ranked second in terms of discharge. It would take the Amazon only about 28 days fill up Lake Erie. The Yangtze, Brahmaputra, Ganges, Yenisy, and Mississippi are other rivers with very high discharges.

The Nile River is more than 4,100 miles long—the longest river in the world. (Jonathan Blair/Corbis)

What is a **delta**?

A delta is a low-lying area where a river meets the sea. Often, the river divides into many distributary streams, forming a triangular-shaped area. The river deposits a large amount of sediment at its mouth, creating excellent soil for farming once the channel of the stream moves. One of the most famous deltas is where the Nile River meets the Mediterranean Sea. Other major deltas include the Mississippi River delta in Louisiana, the Ganges River delta in India, and the Yangtze delta in China. The word delta comes from the Greek letter delta, referring to its triangular shape when written.

What is a **drainage basin**?

The area that includes all of the tributaries for an individual stream or river is its drainage basin. For example, the drainage basin for the St. Lawrence River includes the area surrounding the Great Lakes. Rivers such as the Platte (which has its own drainage basin) flow into the Missouri, and the Missouri flows into the Mississippi. The combined area

Do rivers always flow from north to south?

No, they do not! Rivers always flow from higher ground to lower ground. Though we are familiar with rivers like the Mississippi River in the United States, which flows from north to south, rivers always flow the way gravity takes them. There are many major rivers in Europe, Asia, and North America that flow from south to north, such as the Ob in Russia and the Mackenzie in Canada.

drained by the Platte, the Missouri, the Mississippi, and all other Mississippi River tributaries combined create the third largest drainage basin in the world. The Amazon has the largest drainage basin of any river, the Congo has the second largest.

What is a **tributary**?

Any stream that flows into another stream is a tributary. Most major rivers have hundreds of tributaries, which on a map look like branches of a tree. One classification system of rivers is based upon the number of tributaries a river has.

What is a **watershed**?

A watershed is the boundary between drainage basins. It is usually the crest of a mountain where water flows on either side into two different drainage basins.

What is a **wadi**?

Wadi is the Arabic word for a gully or other stream bed that is dry for most of the year. A wadi is a channel for streams that develop during the short rainy season. The channels of wadis were probably initially carved when the desert regions of today had more rainfall.

What is a **meander**?

Streams and rivers that have carved a flat floodplain commonly flow in curves known as meanders. These S-shaped curves vary by the size and flow of the river. The river flows faster on the outside curve of the meander and therefore continues to cut and create a larger curve.

What is an **oxbow lake**?

An oxbow lake is a crescent-shaped lake that is formed when the meander, or curve of a river, is cut off from the rest of the river during a flood, or when the curve of the meander becomes so large that the river begins flowing along a new path. The curve that remains becomes its own lake. These can commonly be seen along the Mississippi River system.

What is the world's **largest lake**?

The Caspian Sea (which is really a lake) is the largest lake in the world. It is surrounded by Russia, Kazakhstan, Turkmenistan, Iran, and Azerbaijan and is over 143,200 square miles in area. The second largest lake, Lake Superior in North America, is a mere 31,700 square miles.

PRECIPITATION

How is **rainfall measured**?

Agencies like the National Weather Service use very accurate devices that measure rainfall to the nearest one-hundredth of an inch. The devices, known as a rain guages or tipping-bucket gauges, collect rain-

Do oceans get more rain than land?

The oceans receive just over their share, percentage-wise, of the world's precipitation, about 77 percent. The remaining 23 percent of precipitation falls on the continents. Some areas of the world receive far more precipitation than others. Some parts of equatorial South America, Africa, Southeast Asia, and nearby islands receive over 200 inches of rain a year, while some desert areas receive only a fraction of an inch of rain per year.

water, usually at a point unaffected by local buildings or trees that may interfere with the rain.

How can **I measure** how much rain falls where I live?

Any container with a flat bottom and flat sides can measure rainfall. The width of the top of the container must be the same as at the bottom of the container, but the diameter does not matter. It could be a device purchased for measuring precipitation or something as simple as a coffee can.

Where does it **rain the most**?

Mt. Waialeale, on the island Kauai in Hawaii, receives a whopping average of 472 inches of rain a year—that's over 39 *feet* of rain per year!

Where does it **rain the least**?

Northern Sudan's Wadi Halfa (which is in the Sahara Desert) receives an average of less than one-tenth of an inch of rain per year. That's hardly a drop at the bottom of a bucket.

How much **water is in snow**?

When about 10 inches of snow melts, it turns into about one inch of water. Snow has pockets of air between snowflakes when they are on the ground, so it takes 10 times the snow to make an equivalent amount of water.

What is the difference between **snow and hail**?

Snow is water vapor that freezes in clouds before falling to the Earth. Hail is water droplets (raindrops) that have turned to ice in clouds.

How is **hail formed**?

Hail is ice that is formed in large thunderstorm clouds. Hail begins as droplets of water— normally destined to become raindrops—that are blown upward and subsequently freeze. They then fall lower within the cloud, where they collect more water, are blown upward again, and refreeze. The hailstone grows larger as it collects more and more ice, and eventually falls to the ground.

How big was the **largest hailstone**?

The largest recorded hailstone fell in Kansas in 1970 and weighed one pound, 11 ounces. It measured 17.5 inches in diameter.

GLACIERS

What is a **glacier**?

A glacier is a mass of ice that stays frozen throughout the year and flows downhill. They are capable of carving rock with their weight and slow, steady movement. Glaciers are responsible for the stunning landscape

Hayes Glacier moves through a mountain range in Alaska. (USGS/Corbis)

of Yosemite National Park in California. Large glaciers that cover the land are also known as ice sheets.

Are there **still glaciers** in the United States?

Yes, small glaciers exist throughout Alaska, within the Cascade Range of Washington state, sporadically across the Rocky Mountains, and also in the Sierra Nevadas of California.

How **old** are glaciers?

Glaciers present today were created during the last stage of glaciation, the Pleistocene epoch, which lasted from 1.6 million years ago to about 10,000 years ago.

Did glaciers create the **Great Lakes**?

Yes, the Great Lakes are the world's largest lakes formed by glaciers. During the Pleistocene epoch, glaciers inched over the Great Lakes area, moving weak rock out of their way and leaving behind huge, carved basins. As the glaciers began to melt, the basins filled with water and created the Great Lakes.

CONTROLLING WATER

How did the ancient **Romans** get water to their cities?

The ancient Romans and Mesopotamians built aqueducts to transport water between a source and areas where it was needed for agriculture or civilization. The Roman system was very extensive, and was constructed throughout its empire. Some portions of these ancient aqueducts are still in use. Today, modern concrete-lined channels transport water hundreds of miles. The most extensive aqueduct systems in the world

A Roman aqueduct still standing in Italy. (Archive Photos)

today are those that bring water to Southern California from the Colorado River in the east and from the Sacramento River in the north.

What do **dams** do?

By blocking the flow of a river, a dam allows a reservoir of water to build up. Dams are built in order to minimize floods, to provide water for agriculture, and to provide water for recreational uses. Dams in the U.S. are somewhat controversial, as the Bureau of Reclamation and the Army's Corps of Engineers battle to build more dams and control more water in the western U.S. Many outdoor enthusiasts and environmentalists feel that dams are not always necessary.

What is the **tallest dam** in the world?

Tajikistan is home to the world's two tallest dams—Rogun and Nurek. Rogun is about 1100 feet tall (about 100 stories!) and Nurek is about 985 feet tall. The United States' tallest dam, Oroville (in Northern California), is 16th on the world list at 755 feet.

How do **farmers water their crops**?

The process of artificially watering crops is called irrigation. In some areas of the world, agriculture can rely on rainfall for all its water needs. In drier areas (usually those receiving less than 20 inches of rainfall per year), irrigation is required. Water is pumped from aquifers or delivered via an aqueduct to the fields where it flows through small channels between plants, or is sprayed through sprinklers. In very water-conservative regions such as Israel, water is scientifically dripped onto plants, thereby providing the exact amount of water necessary.

CLIMATE

DEFINITIONS

What is the difference between **climate and weather**?

Climate is the long-term (usually 30-year) average weather for a particular place. The weather is the current condition of the atmosphere. So, the weather in Barrow, Alaska, might be a hot 70 degrees Fahrenheit, but their tundra climate is generally polar-like and cold.

How are different types of **climates classified**?

The German climatologist Wladimir Köppen developed a climate classification system that is still used today, albeit with some modifications. He classified climates into six categories: tropical humid, dry, mid-latitude, severe midlatitude, polar, and highland. He also created sub-categories for five of these classifications. His climate map is often found in geography texts and atlases.

What is the **smoggiest city** in the world?

Mexico City, the second most populated city in the world, is also the smoggiest. Mexico City has 24 million people, 40,000 factories, and 3.5 million cars, which serve as the source for most of the city's smog prob-

69

What is global warming?

Global warming is the gradual increase of the Earth's average temperature—which has been rising since the Industrial Revolution. If temperatures continue to increase, some scientists expect major climatic changes, including the rise of ocean levels due to ice melting at the poles. According to many scientists, global warming is primarily due to the greenhouse effect.

lem— a problem that causes disease and death to hundreds of residents annually.

What is a **willy-willy**?

Willy-willy is the Australian name for a hurricane.

THE ATMOSPHERE

How much **pressure** does the atmosphere exert upon us?

Average air pressure is 14.7 pounds per square inch at sea level.

Why is the **sky blue**?

This is one of the world's most frequently pondered questions, and, contrary to what some people believe, the sky's blue color is not due to the reflection of water. Light from the sun is composed of the spectrum of colors. When sunlight strikes the Earth's atmosphere, ultraviolet and blue waves of light are the most easily scattered by particles in the atmosphere. So, other colors of light continue to the Earth while blue

Mexico City, the smoggiest city in the world. (Archive Photos)

and ultraviolet waves remain in the sky. Our eyes can't see ultraviolet light, so the sky appears the only color remaining that we can see, blue.

How many **layers are in the atmosphere**?

There are five layers that make up the Earth's atmosphere. They extend from just above the surface of the Earth to outer space. The layer of the atmosphere that we breathe and exist in is called the troposphere and extends from the ground to about 10 miles above the surface. From about 10 miles to 30 miles up lies the stratosphere. The mesosphere lies from 30 to 50 miles above the surface. A very thick layer, the thermosphere, lies from 50 all the way to 125 miles up. Above the 125 mile mark lies the exosphere and space.

Why can I hear an **AM radio station** from hundreds of miles away at night but not during the day?

At night, AM radio waves bounce off of a layer of the ionosphere, the "F" layer, and can travel hundreds, if not thousands, of miles from their

71

What is the air made of?

The air near the Earth's surface is primarily nitrogen and oxygen—nitrogen comprises 78 percent and oxygen 21 percent. The remaining 1 percent is mostly argon (0.9 percent), a little carbon dioxide (0.035 percent), and other gasses (0.06 percent).

source. During the day, the same reflection of radio waves cannot occur because the "D" layer of the ionosphere is present and it absorbs radio waves.

Why don't **FM radio waves** travel very far?

FM radio waves are "line of site," which means they can only travel as far as their power and the height of their radio antenna will allow. The taller the antenna, the farther the waves can travel along the horizon (as long as they have enough power).

Does **air pressure change** with elevation?

Yes, it does. The higher you go, the less air (or atmosphereic) pressure there is. Air pressure is also involved in weather systems. A low-pressure system is more likely to bring rain and bad weather versus a high-pressure system, which is usually drier. At about 15,000 feet, pressure is half of what it is at sea level.

What are the different **kinds of clouds**?

There are dozens of types of clouds, but they can all be classified into three main categories: cirriform, stratiform, and cumuliform. Cirriform clouds are feathery and wispy; they are made of ice crystals and occur at high elevations. Stratiform clouds are sheet-like and spread out across the sky. Cumuliform clouds are the ubiquitous cloud—puffy and individual, they can be harmless or they can be the source of torrential storms and tornadoes.

How much of the Earth is usually **covered by clouds**?

At any given time, about one-half of the planet is covered by clouds.

How do **airplanes create clouds**?

When the air conditions are right and it's sufficiently moist, the exhaust from airplanes often creates condensation trails, known as contrails. Contrails are narrow lines of clouds that usually evaporate rather quickly. Contrails can turn into cirrus clouds if the air is close to being saturated with water vapor.

What is the **greenhouse effect**?

The greenhouse effect is a natural process of the atmosphere that traps some of the sun's heat near the Earth. The problem with the greenhouse effect, however, is that it has been unnaturally increased, causing more heat to be trapped and the temperature on the planet to rise. The gasses that have caused the greenhouse effect were added to the atmosphere as a byproduct of human activities, especially combustion from automobiles.

What is **albedo**?

Albedo is the amount of the sun's energy that is reflected back from the surface of the Earth. Overall, about 33 percent of the sun's energy bounces off the Earth and its atmosphere and travels back into space. Albedo is usually expressed as a percentage.

What is the **jet stream**?

The jet stream is a band of swiftly moving air located high in the atmosphere. The jet stream meanders across the troposphere and stratos-

A stream of clouds, pushed by a jet stream, moves over the Middle East. (NASA/Corbis)

phere (up to 30 miles high) and affects the movement of storms and air masses closer to the ground.

OZONE

What is the **ozone layer**?

The ozone layer is part of the stratosphere, a layer of the Earth's atmosphere that lies about 10 to 30 miles above the surface of the Earth.

Ozone is very important to life on the planet because it shields us from most of the damaging ultraviolet radiation from the sun.

Is the ozone layer being **depleted**?

Scientists have recognized that a hole has developed in the ozone layer, a hole that has been growing since 1979. The hole is located over Antarctica and has been responsible for increased ultraviolet radiation levels in Antarctica, Australia, and New Zealand. As the ozone hole grows, it will increase the amount of harmful ultraviolet light reaching the Earth, causing cancer and eye damage, and killing crops and micro-organisms in the ocean.

How does ozone **disappear**?

Chlorofluorocarbons (CFCs), which have been used as refrigerants in air conditioners and refrigerators, and halon, which is used in fire extin-guishing systems, rise up in the atmosphere to the ozone layer. Ultravi-olet rays break the CFCs into bromine and chlorine, which destroy ozone molecules.

CLIMATIC TRENDS

Why is it **very wet** on one side of a mountain range?

It's much more wet on one side of a mountain than the other because of a process known as orographic precipitation. Orographic precipitation causes air to rise up the side of a mountain range and cool, creating storms. The storms deposit a great deal of precipitation on that side of the mountain and create a rain shadow effect on the opposite side of the range. The Sierra Nevada mountains are an excellent example of oro-graphic precipitation because the mountains of the western Sierras receive considerable rainfall (far more than California's Central Valley), while the eastern Sierras of Nevada are quite dry.

What is a **rain shadow**?

When the moisture in the air is squeezed out by orographic precipitation, there's not much left for the other side of the mountains. The dry side of the mountain experiences a rain shadow effect because they are in the shadow of the rain.

What is **El Niño**?

El Niño (also known as ENSO or the El Niño Southern Oscillation), is a large patch of warm water that moves between the eastern and western Pacific Ocean near the equator. When the warm water (about one degree Celsius warmer than normal) of El Niño is near South America, the warm water affects the weather in the southwestern U.S. by increasing rainfall, and is responsible for changes in the weather throughout the world. El Niño lasts for about four years in the eastern Pacific Ocean and then returns to the western Pacific near Indonesia for another four years. When the warm water is in the western Pacific, it is known as La Niña, the opposite of El Niño. When La Niña is in action, we have "normal" climatic conditions.

Where does the **name El Niño** come from?

The phenomenon of El Niño was discovered by Peruvian fishermen who noticed an abundance of exotic species that arrived with the warmer water. Since this usually occurred around the Christmas season, they called the phenomenon El Niño, which means "the baby boy" in Spanish, in honor of the Christ child. La Niña, the opposite cycle of El Niño, means "the baby girl."

What are **ice core samples** and why are they important?

An ice core sample is a thick column of ice, sometimes hundreds of feet long, that is produced by drilling a circular pipe-like device into thick ice and then pulling out the cylindrical piece. Ice core samples from places like Greenland and Antarctica provide scientists with important clues about past climates. Air trapped in the ice remains there for thousands of

Can people live in the torrid zone?

The ancient Greeks divided the world into climatic zones that are not accurate. The three zones included frigid, temperate, and torrid. They believed that civilized people could only live in the temperate zone (which, of course, was centered around Greece). From Europe northward was part of the inhospitable frigid zone, while most of Africa was torrid. Unfortunately, this three-zone classification system stuck and was later expanded to five zones once the southern hemisphere was explored. People identify everything north of the Arctic Circle (near northern Russia) and south of the Antarctic Circle (near the coast of Antarctica) as frigid, everything between the tropics and the Arctic and Antarctic circles as temperate, and the zone between the Tropic of Cancer and Capricorn as torrid.

years, so when scientists collect ice cores they can analyze the air to determine the composition of the atmosphere at the time the ice was formed. Sediments and tiny bugs are also found in the ice and provide additional clues to the state of the natural world at the time the ice was deposited.

What is **continentality**?

Areas of a continent that are distant from an ocean (such as the central United States) experience greater extremes in temperature than do places that are closer to an ocean. These inland areas experience continentality. It might be very hot during the summer, but it can also get very cold in winter. Areas close to oceans experience moderating effects from the ocean that reduce the range in temperatures.

How does land **turn into desert**?

The process known as desertification is complicated and results from such activities as overgrazing, inefficient irrigation systems, and deforestation. It is most widespread in the Sahel region of Africa, a strip of land along the southern margin of the Sahara desert. The Sahara grows larger because of desertification. Desertification can be reversed by changing agricultural practices and by replanting forests.

WEATHER

How do I **convert** Fahrenheit to Celsius to Kelvin?

Fahrenheit and Celsius are two common temperature scales used throughout the world. Temperature in Fahrenheit can be converted to Celsius by subtracting 32 and multiplying by five; divide that number by nine and you have Celsius. Conversely, you can convert Celsius to Farenheit by adding 32, multiplying by nine and finally dividing by five. Kelvin, a system used by scientists, is based on the same scale as Celsius. All you have to do is add 273 to your Celsius temperature to obtain Kelvin. Zero degrees Kelvin is negative 273 degrees Celsius.

What is a **low high** temperature and a **high low** temperature?

When meteorologists look at daily temperature, there is always a low and a high temperature for each day. If the high temperature is the coldest high temperature for that day or for the month, you have a new record—a new low high. Conversely, if the low temperature for a day is quite warm and breaks records, that's a new high low!

Why is it more likely to **rain in a city** during the week than on the weekend?

Urban areas have an increased likelihood of precipitation during the work week because intense activity from factories and vehicles produce particles that allow moisture in the atmosphere to form raindrops. These same culprits also produce warm air that rises to create precipitation. A study of the city of Paris found that precipitation increased throughout the week and dropped sharply on Saturday and Sunday.

What does a **40 percent chance of rain** really mean?

When the morning weather report speaks of a 40 percent chance of rain, it means that throughout the area (usually the metropolitan area) there is a 4 in 10 chance that at least .001 of an inch of rain will fall on any given point in the area.

Why is it **hotter in the city** than in the countryside?

Cities have higher temperatures due to an effect known as the urban heat island. The extensive pavement, buildings, machinery, pollution

79

from automobiles, and other things urban cause this warmth in the city. Cities such as Los Angeles can be up to five degrees hotter than surrounding areas due to the urban heat island effect. The term comes from temperature maps of cities where the hotter, urban areas look like islands when isotherms (lines of equal temperature) are drawn.

What is a **thunderstorm**?

Thunderstorms are localized atmospheric phenomena that produce heavy rain, thunder and lightening, and sometimes hail. They are formed in cumulonimbus clouds (big and bulbous) that rise many miles into the sky. Most of the southeastern United States has over 40 days of thunderstorm activity each year, and there are about 100,000 thunderstorms across the country annually. Thunderstorms are different from typical rainstorms because of their lightening, thunder, and occasional hail.

What is **air pollution**?

Air pollution is caused by many sources. There are natural pollutants that have been around as long as the Earth, such as dust, smoke, volcanic ash, and pollens. Humans have added to air pollution with chemicals and particulates due to combustion and industrial activity.

WIND

Where does **wind come from**?

The Earth's atmospheric pressure varies at different places and times. Wind is simply caused by the movement of air from areas of higher pressure to areas of lower pressure. The greater the difference in pressure, the faster the wind blows. Some detailed weather maps show wind speed along with isobars (areas of equal air pressure) indicating the level of air pressure.

Pollution rising from factory smokestacks. (UPI/Corbis-Bettmann)

What are the **westerlies**?

These winds flow at midlatitudes (30 to 60 degrees north and south of the equator) from west to east around the Earth. The high-altitude winds known as the jet stream are also westerlies.

What are **monsoons**?

Occurring in southern Asia, monsoons are winds that flow from the ocean to the continent during the summer and from the continent to the ocean in the winter. The winds come from the southwest from April to October, and from the the northeast (the opposite direction) from October to April. The summer monsoons bring a great deal of moisture to the land. They cause deadly floods in low-lying river valleys, but also provide the water southern Asia relies upon for agriculture.

What are **dust devils**?

These columns of brown, dust-filled air, which can rise dozens of feet, are not as evil as the name suggests. They are caused by warm air rising

81

In which direction does the west wind blow?

It blows from the west to the east. Wind is named after the direction from whence it comes.

on dry, clear days. Winds associated with dust devils can reach up to 60 miles per hour and cause some damage, but they are not as destructive as tornadoes and usually die out pretty quickly.

Is Chicago really the **"windy city?"**

Chicago is not the windiest city in the United States. Chicago's average wind speed of 10.4 miles per hour is beat by Boston (12.5), Honolulu (11.3), Dallas and Kansas City (both 10.7), and especially in the true windy city, Mt. Washington, New Hampshire (35.3).

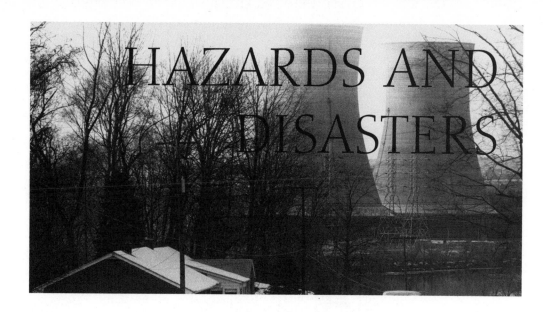

HAZARDS AND DISASTERS

What is a **hazard**?

A hazard is any source of danger that can cause injury or death to humans or that can cause property damage. Hazards range from airline accidents to tsunamis to asteroids smashing into the Earth.

Why is it important to have an **out-of-state contact** in case of a disaster?

It's usually easier to call outside of a disaster area than inside one. By identifying a relative or friend who lives outside of your home state as an emergency contact, your family can ensure communication following a disaster.

What is the difference between a **watch and a warning**?

The U.S. National Weather Service issues watches and warnings for a variety of hazards when they may be imminent. A watch (such as a tornado watch or a flood watch) means that such an event is likely to occur or is predicted to occur. A warning is more serious. It means that a hazard is already occurring or is imminent. Warnings are usually broadcast on television and radio stations via the Emergency Alert System (formerly known as the Emergency Broadcast System).

83

What's the difference between the old Emergency Broadcast System and the Emergency Alert System?

The Emergency Broadcast System (EBS), created in 1964 to warn the country of a national emergency such as nuclear attack, became the Emergency Alert Service (EAS) in 1997. The old EBS system relied on one primary radio station in each region to receive an emergency message and then broadcast it to the public and other media outlets. The new system, which also includes cable television, operates via computer and can be automatically and immediately broadcast to the public. It also allows additional local governmental agencies the opportunity to broadcast emergency messages. Future plans for the EAS include radios and televisions that will automatically turn on when an alert is announced.

How should we **prepare for disaster**?

Disasters can and do happen everywhere. You should prepare for disaster by having a disaster supply kit with supplies for you and everyone in your family available at home and work, as well as a mini kit in your automobile. It should include food, water, first aid equipment, sturdy shoes, an AM/FM radio (with batteries kept outside of the radio), a flashlight (with batteries kept outside of the flashlight), vital medication (especially prescription medication), blankets, cash (if the power and computers are down, credit and ATM cards won't work), games and toys for children, and any other essentials. Contact your local chapter of the Red Cross for more information about disaster preparedness.

Should we use **candles** after a disaster or power outage?

Many deaths and a great deal of property damage have been caused by fires resulting from people using candles following a disaster. People leave candles burning as a source of light, but these can fall over and start fires. It is strongly advised that people not use candles when the

The Red Cross brings disaster relief to an Ethiopian village in 1985. (Chris Rainier/Corbis)

power goes out. There are many flashlights and battery operated lanterns that are available commercially and should be part of your disaster supply kit.

What is the leading cause of **disaster-related death** in the United States?

Lightening is the leading cause of disaster-related death in America. From 1940 to 1981, about 7,700 people died from lightening strikes, 5,300 from tornadoes, 4,500 from floods, and 2,000 from hurricanes. So, it's best to avoid open spaces, high ground, water, tall metal objects, and metal fences during an electrical storm.

How can I learn more about **disasters in my town**?

Each community should have its own disaster plan that includes a history of past disasters (those that have happened in the past are likely to occur in the future) along with plans for dealing with future disasters.

85

What do medical geographers do?

Medical geographers and epidemiologists (scientists who study disease and health) regularly use maps and spatial information to help control illness and death. Mapping has solved the mystery behind high levels of cancer in small areas and has been used to understand the spread of AIDS. Medical geographers don't just study the distribution of disease, they also investigate the accessibility of people to health services.

You should be able to consult this plan to learn how your community would cope with disaster and to find out the locations of evacuation routes and shelters. Many communities place important disaster planning information in the telephone book for easy reference.

What is the best way to **help after a disaster**?

Disaster relief agencies such as the Red Cross are in vital need of money after a disaster to purchase necessary items for victims or provide financial support to them. Call your local chapter of the Red Cross to find out how to help. Donating food or clothing places additional burdens on the agencies in the immediate aftermath of a disaster, as personnel are not available to sort, clean, or distribute donated goods.

How did a map help stop the spread of **cholera**?

During an 1854 cholera outbreak in London, a physician named John Show mapped the distribution of cholera deaths. His map showed that there was a high concentration of deaths in an area surrounding one specific water pump (water had to be hand-pumped and carried in buckets at the time). When the handle was taken off of the water pump, the number of cholera deaths plummeted. When it was determined that cholera could be spread through water, future epidemics were curbed. This was the beginning of medical geography.

Which natural disasters doesn't **Southern California** experience?

While urban southern California is plagued by earthquakes, wildland fires, floods, landslides, and tornadoes (yes, even tornadoes), they rarely receive snowstorms or hurricanes.

What is the **Ring of Fire**?

If you were to look at a map of the world's major earthquakes and volcanoes, you would notice a pattern circling the Pacific Ocean. This dense accumulation of earthquakes and volcanoes is known as the Ring of Fire. The ring is due to plate tectonics and the merger of the Pacific plate with other surrounding plates, which creates faults and seismic activity (especially Alaska, Japan, Oceania, and the west coasts of North and South America), along with volcanic mountain ranges such as the Cascades of the U.S. Pacific Northwest and the Andes of South America.

VOLCANOES

What are **volcanoes**?

Volcanoes are the result of magma rising or being pushed to the surface of the Earth. Hot liquid magma, which is located under the surface of the Earth, rises through cracks and weak sections of rock. The mountain surrounding a volcano is formed by lava (called magma until it arrives at the Earth's surface) that cools and hardens, making the volcano taller or wider or both.

How many **active volcanoes** are there in the world?

There are about 1,500 active volcanoes around the world. Most are located in the Ring of Fire surrounding the Pacific Ocean. About one-tenth of the world's active volcanoes are located in the United States. A volcano is considered active if it has erupted in the last 10,000 years.

The ruins of Pompeii, with Mount Vesuvius in the background. (Archive Photos)

Where are the **active volcanoes in the United States**?

Washington, Oregon, and California have many potentially active volcanoes. The most recent eruption in the United States was that of Mount St. Helens in southern Washington state in 1980. Other volcanoes in the region, such as Mount Shasta, Lassen, Rainier, and Hood could erupt with little warning.

What is the difference between **magma and lava**?

Magma is hot, liquefied rock that lies underneath the surface of the Earth. When magma erupts or flows from a volcano onto the Earth's surface, it becomes lava. There is no difference in substance, only the name changes.

How was **Pompeii** destroyed?

In the year 79, the volcano Mount Vesuvius erupted and buried the ancient Roman town of Pompeii under 20 feet of lava and ash. Neighboring Hurculaneum was also destroyed. Pompeii is famous because excavations of the city, which began in 1748 and continue to this day, provide an excellent look at Roman life at the beginning of the millennium. The covering of the city by debris preserved not only the places where people last stood but also paintings, art, and many other artifacts.

EARTHQUAKES

What causes **earthquakes**?

The tectonic plates of the Earth are always in motion. Plates that lie side by side may not move very easily with respect to one another; they "stick" together and occasionally they slip. These slips (from a few inches to many feet) create earthquakes and can often be very destructive to human lives and structures.

What is an **epicenter**?

An epicenter is the point on the Earth's surface that is directly above the hypocenter, or point where earthquakes actually occur. Earthquakes do not usually occur at the surface of the Earth but at some depth below the surface.

What is a **fault**?

A fault is a fracture or a collection of fractures in the Earth's surface where movement has occurred. Most faults are inactive, but some, like California's San Andreas fault, are quite active. Geologists have not discovered all of the Earth's faults, and sometimes earthquakes occur that take the world by surprise, like the one in Northridge, California, in 1994. When earthquakes occur on faults that were previously unknown, they are called blind faults.

89

What is the significance of the **San Andreas fault**?

The infamous San Andreas fault is the border between the North American and the Pacific tectonic plates. This fault lies in California and is responsible for some of the major earthquakes that occur there. Los Angeles is on the Pacific Plate but San Francisco is on the North American Place. The Pacific Plate is sliding northward with respect to the North American Plate and, as a result, Los Angeles gets about half an inch closer to San Francisco every year. In a few million years, the two cities will be neighbors.

Was **San Francisco** destroyed by **earthquake** or by fire in 1906?

In 1906 a very powerful earthquake struck San Francisco, California, which sparked a fire that destroyed much of the city. In an effort to preserve San Francisco's image with residents and would-be visitors, official policy regarding the disaster stated that it was not the earthquake but mostly the fire that destroyed the city. Official books and publications produced after the earthquake referred to both the fire and the earthquake as having caused the damage. In fact, the earthquake did considerable damage to the city and killed hundreds.

Which states are **earthquake-free**?

While a 20-year period isn't an excellent indicator, there were four states that had no earthquakes between 1975 and 1995: Florida, Iowa, North Dakota, and Wisconsin.

Is there a high risk of **earthquakes in the Midwestern U.S.**?

Great earthquakes struck the New Madrid, Missouri, area in 1811 and 1812. They caused considerable damage (some areas experienced shaking at the level of XI on the Mercalli scale) and were felt as far away as the East Coast. The potential exists for future earthquakes in the region, since earthquakes have there occurred before. Planning and preparedness continues throughout the region, centered at the

The wreakage of an earthquake-damaged expressway in Kobe, Japan, 1995. (Reuters/Archive Photos)

junction of Missouri, Arkansas, Illinois, Kentucky, Tennessee, and Mississippi.

What should I do **in the event of an earthquake**?

Duck, cover, and hold! Duck under a table, counter, or any area that can provide protection from falling objects. Cover the back of your head with your hands to help protect against flying debris. Hold on to the leg of the table or anything solid to ride out the shaking.

Is it safe to **stand in a doorway** during an earthquake?

While a doorway is a nice, structurally sound place to be during an earthquake, officials have found that many people are injured when a door swings open and closed during an earthquake, so you may want to avoid standing in a place where your fingers can become crushed.

What is the **Richter scale**?

The Richter scale measures the energy released by an earthquake. It was developed in 1935 by California seismologist Charles F. Richter.

With each increase in Richter magnitude, there is an increase of 30 times the energy released by an earthquake. For example, a 7.0 earthquake has 30 times the power of a 6.0, and an 8.0 is 900 times as powerful as a 6.0. Each earthquake only has one Richter magnitude. The strongest earthquakes are in the 8.0 range—8.6 for Alaska's 1964 earthquake, and 8.0 for China's 1976 earthquake in Tangshan.

What is the **Mercalli scale**?

The Mercalli scale measures the power of an earthquake as felt by humans and structures. It was developed in 1902 by Italian geologist Giuseppe Mercalli. The Mercalli scale is written in roman numerals and it ranges from I (barely felt) to XII (catastrophic). The Mercalli scale can be mapped surrounding an epicenter and will vary based on the geology of an area.

The Mercalli Scale of Earthquake Intensity

I	Barely felt
II	Felt by a few people, some suspended objects may swing
III	Slightly felt indoors as though a large truck were passing
IV	Felt indoors by many people, most suspended objects swing, windows and dishes rattle, standing autos rock
V	Felt by almost everyone, sleeping people are awakened, dishes and windows break
VI	Felt by everyone, some are frightened and run outside, some chimneys break, some furniture moves, causes slight damage
VII	Considerable damage in poorly built structures, felt by people driving, most are frightened and run outside
VIII	Slight damage to well-built structures, poorly built structures are heavily damaged, walls, chimneys, monuments fall
IX	Underground pipes break, foundations of buildings are damaged and buildings shift off foundations, considerable damage to well-built structures
X	Few structures survive, most foundations destroyed, water moved out of banks of rivers and lakes, avalanches and rockslides, railroads are bent
XI	Few structures remain standing, total panic, large cracks in the ground
XII	Total destruction, objects thrown into the air, the land appears to be liquid and is visibly rolling like waves

How many **really big earthquakes** occur each year?

On average, there are about 100 earthquakes of magnitude 6.0–6.9, about 20 of magnitude 7.0–7.9, and two huge 8.0–8.9 earthquakes each year. Many of these really big earthquakes occur in the ocean, so we don't hear much about them.

Is a **magnitude 10** the top of the Richter scale?

While the media often refers to the Richter scale as being on a scale of 1 to 10, there is no upper limit, even though the strongest quakes are not as high as 10. It is incorrect to assume that a 7 is on a scale of 1 to 10 because the magnitudes are based on the energy released and it is a logarithmic scale.

TSUNAMIS

What is a **tsunami**?

A tsunami, also known as a seismic sea wave, is usually caused by an earthquake that occurs under the ocean or near the coast. The seismic energy creates a large wave that can cause heavy damage hundreds or even thousands of miles from its source. Hawaii is frequently struck by tsunamis.

How does **Hawaii** protect itself from tsunamis?

There is a sophisticated global monitoring network that provides warnings about possible tsunamis, allowing the islands of Hawaii and other coastal areas to prepare for impending disaster. Hawaii also has a thorough evacuation system to protect lives in the face of tsunami danger.

How are hurricanes ranked?

Hurricanes are ranked on a scale of one to five, with category one hurricanes being the weakest and category five being the strongest and most destructive. The damage caused by each category of hurricane ranks from: 1, minimal; 2, moderate; 3, extensive; 4, extreme (such as Hurricane Andrew in 1992); to 5, catastrophic.

HURRICANES

What part of a hurricane is **most damaging**?

Floods caused by hurricanes are the most destructive element. The low-pressure center of a hurricane causes a mound of water to rise above the surrounding water. This hill of water is pushed by the hurricane's fierce winds and low pressure onto the land, where it floods coastal communities, causing significant damage. Hurricanes also spark tornadoes that contribute to the devastation.

How fast do **hurricane winds** blow?

The strongest hurricanes have winds that reach speeds well over 150 miles per hour.

FLOODS

How much rain does it take to make a flood?

The amount varies widely for different areas. In some U.S. western deserts, or in some large urban areas, just a few minutes of strong rain

Why do people live in floodplains?

People have lived in floodplains for thousands of years. Fertile land for agriculture lines the floodplain, and the nearby water source makes life easier. Unfortunately, when the river does flood, these communities are severely damaged and people suffer. Hazard mitigation, such as levees, dams, dikes and other structures, attempt to limit damage during floods. Sometimes, when the structures fail (such as a levee breaking), large areas are inundated with water. Inhabitants of floodplains must balance the risks with the rewards of living in such an unpredictable environment.

will cause a flash flood in canyons and low-lying areas. In areas prone to greater rainfall amounts, it often takes quite a bit more rain (sometimes a few days' or weeks' worth) to cause rivers to overflow and dams to fill up, raising concerns of those who live downstream. Areas that normally receive more rainfall have better natural drainage systems and are usually home to plants that readily absorb the extra water.

What have been some of the **most destructive floods** in history?

In the United States, the failure of a dam in 1889 upstream from the community of Johnstown, Pennsylvania, killed 2,200 people. Some of the world's most catastrophic flooding takes place in China. A flood on the Huange He River in 1931 killed 3.7 million people.

What is a **floodplain**?

A floodplain is the area surrounding a river that, when unmodified by human structures, would normally be flooded during a river flood. A floodplain can be a few feet or many miles wide, depending on the river flow as well at the local terrain. Even though levees and flood walls can be built (with homes and businesses built just behind them), the floodplain does not vanish. If the structures break or are damaged, the water from a flood can fill a floodplain, just as it did before humans occupied it.

What is a **hundred-year flood**?

A hundred-year flood refers not only to the size of a flood, but also to the odds of it occurring. A hundred-year flood has a one percent (or one in a hundred) chance of occurring in any given year. It has no relationship to the frequency of occurrence. The magnitude of such a flood is relative to the frequency of occurrence, so a 100-year flood is much larger than any run-of-the-mill annual flood. A 500-year flood only has a one in 500 (0.2 percent) chance of occurring in any given year and would be much larger and more devastating than a 100-year flood.

What is the **National Flood Insurance Program**?

The National Flood Insurance Program (NFIP) was established by the U.S. Federal Government in 1956 as a subsidized insurance program for home and business owners. The government began the program by creating Flood Insurance Rate Maps (FIRM) showing the boundaries of 100-year and 500-year flood zones. The cost of the insurance is based on the flood risk. The Federal Emergency Management Agency (FEMA) oversees the program and requires the purchase of flood insurance by any owner affected by a disaster before they can be provided with disaster assistance. This way, the next time a flood occurs, they will be insured.

How can I obtain a **flood map** of my community?

The best way to see a Flood Insurance Rate Map (FIRM) for your area would be to contact your local government. Their planning or emergency management agency should have the FIRM maps available. Purchasing them from FEMA is not recommended because the maps change often and are best interpreted by a planning or emergency expert.

What should I do in the event of a **flood**?

If a flood is expected, turn on your battery-powered radio and listen for information about when and where to evacuate. If a flood or flash flood is

97

coming toward you, move quickly to a higher elevation—but don't ever try to outrun a flood. Also, don't drive through standing water, as it can quickly rise and stall your vehicle, possibly trapping you among swift water.

TORNDOES

What are **tornadoes**?

Tornadoes are very powerful, yet tiny storms that have destructive winds capable of leveling buildings and other structures. Winds in a tornado form a dark gray column of air, with the center of the tornado acting like a vacuum, picking up objects and moving them along the storm's path. Tornadoes can last from a few minutes to an hour.

What should I do when a **tornado** approaches?

Try to get to the lowest level of the building (unless you are in a mobile home or outdoors, in which case you should seek a sturdy and safe shelter). Go to the center of the room and hide under a sturdy piece of furniture. Stay away from windows, hold on to the leg of a table or something else stable, and protect your head and neck with your arms.

Where is **tornado alley**?

Tornadoes occur more frequently in the central United States than anywhere else in the world. Tornado Alley is an area stretching from northwest Texas, across Oklahoma (the tornado capital of the world), and through northeast Kansas. On average, over 200 tornadoes occur across Tornado Alley each year.

What is the **most dangerous state** to live in due to tornadoes?

Massachusetts is considered the most dangerous state to live in due to tornadoes. While Oklahoma receives far more tornadoes than Massachu-

Tornadoes are very powerful, yet tiny storms that have destructive winds capable of leveling buildings and other structures. (Archive Photos)

setts does, the population density and risk of death or severe injury is greater in the New England state.

What is the **Fujita scale** of tornado intensity?

The Fujita scale measures the strength of a tornado based on observed damage and effects. The scale ranges from F0 (a weak tornado) through F6 (an almost inconceivable tornado, having close to no chance of actually occurring). About 3/4 of all tornadoes are weak (F0–F1), while only one percent are violent (F4–F5).

Are there **tornadoes in Europe**?

While 90 percent of all tornadoes occur in the United States, there are tornadoes in Europe, especially in western France. Other tornado regions of the world include eastern and western Australia, southern Brazil, Bangladesh, South Africa, and Japan.

Three Mile Island, site of the worst nulcear accident in U.S. history. (David H. Wells/Corbis)

OTHER HAZARDS AND DISASTERS

What is **acid rain**?

Motor vehicles and industrial activity release tons of pollutants into the air. When mixed together, the pollutants form sulfuric and nitric acids that later fall to the ground in rain or snow. This precipitation is known as acid rain. Acid rain is responsible for damaging lakes by killing plant and animal life and for killing trees around the world. Canada has been especially hard-hit by acid rain caused by industrial activities in the United States.

Does **radiation from a nuclear plant** stop at the 10-mile zone?

U.S. nuclear power plants are required to create emergency planning zones within a 10-mile radius surrounding their plants. These imaginary 10-mile lines are not walls that hold back the effects of radiation, but simply a distance determined by emergency planners. In the event of an accident, the residents of the 10-mile zone might not need to be evacuated but could be advised to remain indoors with their windows closed. Nuclear plants also establish smaller zones of two and five miles surrounding the plants, within which the risk of radiation exposure is much greater.

What happened at **Three Mile Island**?

Three Mile Island, Pennsylvania, was the site of the United States' worst nuclear accident. Luckily, no radiation was released into the environment and no one was killed. In March, 1979, the nuclear reactor at the Three Mile Island plant overheated, breaking the radioactive rods. Pennsylvania's governor recommended a voluntary evacuation of pregnant women and preschool children who lived within five miles of the plant. It was the unexpected self-evacuation of residents in the area that created major problems. The evacuations yielded surprising information about the lack of preparedness of communities for such an event, and have led to increased planning and preparedness for nuclear accidents and evacuations.

What is **nuclear winter**?

A nuclear winter is what would follow a large-scale nuclear war. Radioactive particles, dust, and smoke released into the atmosphere would create a large cloud over the planet, blocking out sunlight and reducing temperatures worldwide. Plants and animals would die due to the extremely low temperatures. An extended nuclear winter could cause the death of millions of people from starvation, cold, and other problems.

What caused the **Bhopal disaster**?

In December, 1984, the U.S.–owned Union Carbide pesticide plant in Bhopal, India, leaked toxic chemicals (methyl isocyanate gas) that killed over 3,800 people. It was the worst industrial accident in history. Union Carbide paid a fine of $470 million to avoid facing criminal charges.

101

Does lightening ever **strike twice** in the same place?

Lightening can and often does strike in the same place twice. Since lightening bolts head for the highest and most conductive point, that point often receives multiple strikes of lightening in the course of a storm—so stay away from something that has already been struck by lightening! Tall buildings (such as the Empire State Building) often receive numerous lightening strikes during a storm.

TRANSPORTATION AND URBAN GEOGRAPHY

URBAN SPRAWL

CITIES AND SUBURBS

What is a **city**?

In the United States, a city is a legal entity, delegated power by a state and county to govern and provide services to its citizens. Cities also have charters, which are somewhat akin to local constitutions, and have specific boundaries.

What was the first city to have more than **one million people**?

During ancient times, Rome was the world's first city to have a population larger than one million. Rome's population declined during the fall of the Roman Empire in the fifth century, and a city with a population of one million wasn't again seen until the early 19th century in London.

What is an **urban area**?

An urban area consists of a central city and its surrounding suburbs. Urban areas are also known as metropolitan areas. In some cases, urban areas can spread dozens of miles beyond the central city.

An aerial view of Tokyo, Japan, the most populated urban area in the world. (Yann Arthus-Bertrand/Corbis)

What is the **most populated urban area** in the world?

The Tokyo, Japan, urban area is the world's most populated, with over 27 million people (more than one-fifth of Japan's total population). Tokyo lies on Japan's flat Kanto Plain and surrounds Tokyo Bay. Other large urban areas include Mexico City, Mexico (24 million); Sao Paulo, Brazil (22 million); New York City, United States (20 million); Mumbai (formerly Bombay), India (16.6 million); and Shanghai, China (16 million).

What are the **largest metropolitan areas** in the United States?

The New York metropolitan area is America's largest, with 20 million people. Los Angeles is second with 15 million; Chicago is third with 8.5 million; Washington D.C. is fourth with 7 million; and San Francisco is fifth with 6.5 million.

What is a **megalopolis**?

Geographer Jean Gottmann developed the term megalopolis to describe
the huge, interconnected metropolitan area from Boston to Washing-

ton, D.C. "Boswash," as this original megalopolis has been called, has been joined by such nascent megalopoli as "Chi-pitts" (from Chicago to Pittsburgh), the Ruhr area in Germany, Italy's Po Valley, and "San-San" (from San Francisco to San Diego).

What is a **central business district**?

A central business district (CBD) of a city is located downtown, often where the city began, and is the primary concentration of commercial buildings.

Who decides where **houses can be built**?

Almost all American city governments have a department that is responsible for planning the layout of the city. The planning department for each city enforces and delineates zoning, which regulates the location of homes, businesses, factories, and even nuclear power plants.

SUBURBS

When did **suburbs** become fashionable?

Following World War II, the subsequent housing boom and construction of interstate highways led to the development of low-density housing surrounding cities. These areas of low-density housing are known as suburbs. Suburbs have been extremely popular since the 1950s.

What was **Levittown**?

The three Levittowns were large housing developments built from the mid-1940s through the early 1960s by William J. Levitt and his construction company. Levitt invented a process to mass-produce homes by making each home exactly the same. The first Levittown was located in New York and consisted of 17,000 homes. The subsequent Levittowns were built in New Jersey and Pennsylvania. Levittowns were the forerunner of the suburb.

URBAN STRUCTURES

What is the **largest enclosed building** in the world?

Hong Kong's Container Freight Station is the largest enclosed structure in the world, with over seven million square feet of enclosed space.

What is the **largest office building** in the world?

The United States is home to the largest office building in the world, the Pentagon, which has 3.7 million square feet of space under its roof. The Pentagon is home to the country's Department of Defense.

What is the **tallest structure** in the world?

The CN Tower in Toronto, Canada, built as a television transmission tower, is the world's tallest self-supporting structure, at 1,815 feet.

What was the world's first skyscraper?

Completed in 1885 in Chicago, Illinois, the Home Insurance Company Building was the world's first skyscraper.

What is the **tallest building** in the world?

The twin Petronas Towers in Kuala Lumpur, Malaysia, each 1,483 feet and 88 stories tall, are the tallest buildings in the world. Though the Sears Tower in Chicago, Illinois, has more stories (110), it is the second-tallest building in the world at 1,450 feet.

TRANSPORTATION: AIR, LAND, AND SEA

FLIGHT

What was the **first airplane** flown?

Orville and Wilbur Wright were the first two people to fly in a heavier-than-air vehicle, called the "Flyer." They made their historic flight on December 17, 1903, at Kitty Hawk, North Carolina.

What is the **busiest airport** in the world?

Over 69 million people arrive and depart annually through O'Hare Airport in Chicago, Illinois, making it the busiest airport in the world. O'Hare has about 6 million more passengers each year than Hartsfield International Airport in Atlanta, Georgia, the world's second-busiest airport. Dallas–Ft. Worth Airport in Texas is the third-busiest (58 million), followed closely by Los Angeles International in California (57.9 million).

What is the **busiest airport outside of the United States**?

Heathrow Airport in London, England, with over 56 million people arriving and departing each year, is the busiest airport outside the United States, making it the fifth-busiest in the world.

ROADS AND RAILWAYS

Do all roads really lead to **Rome**?

Not any more. During the time of the Roman Empire, the Romans built a massive road network to ensure easy travel in all weather conditions between Rome and the furthest reaches of the Empire. The Romans made their roads as straight as possible and paved large sections of them by precisely piecing together cut rock to make a flat surface. Along the 50,000 miles of Roman roads, markers were placed every Roman mile (just short of a modern mile) so as to indicate either the distance to Rome or to the city where the road originated. After the fall of Rome, the maintenance of the Roman road system was severely neglected, and during the Middle Ages the roads became overused and dilapidated. Though the Romans built these roads over 2,000 years ago, some segments are still in use today.

What is a **turnpike**?

A turnpike is a toll road. In the late 18th century, private companies in the United States and in the United Kingdom built roads and charged users to pass. Beginning in the 1840s, turnpikes had to compete for traffic, and thus profits, with the railroads. The name turnpike is still common on toll highways in the eastern United States, such as the New Jersey Turnpike, the Massachusetts Turnpike, and the Pennsylvania Turnpike.

Freeways like this one help make up the 2,335,000 miles of paved road in the United States. (Archive Photos)

What road in the United States was known as the **National Road**?

The Cumberland Road, also known as the National Road, was the first federally funded road in the United States. Though construction began in 1811, the Cumberland Road was not completed until 1852. Stretching 800 miles from Cumberland, Maryland, to Vandalia, Illinois, the road was built to allow settlers to traverse the Appalachian Mountains and settle in the West. With the advent of the automobile, the road was paved, and in 1926 became part of U.S. Route 40, which stretches across the continent.

What do the **Cumberland Road and Cumberland Gap** have to do with each other?

Absolutely nothing. The Cumberland Road is more than 100 miles from the Cumberland Gap. The Cumberland Gap, which lies near the border of Kentucky, Tennessee, and Virginia, is a pass through the Appalachian Mountains at the Cumberland Plateau. The name "Cumberland" was extremely popular in Colonial America, originating in the name of the British Duke of Cumberland.

How are interstate highways numbered?

One- and two-digit interstate highways are numbered according to their direction. Highways that run in an east-west direction are even numbered, while highways that run in a north-south direction are odd numbered. The lowest numbers are in the south and west, while higher numbers are in the north and east. For example, Interstate 10 is an east-west highway that runs from Santa Monica, California, to Jacksonville, Florida; thus, it has a even, low number. Interstate 95 is a north-south highway that runs from Houlton, Maine, to Miami, Florida; thus, it has an odd, high number. Three-digit interstate highways are short spur routes connected to a two-digit interstate.

What is the difference between a **highway and a freeway**?

The term highway can be used for any road, but most often describes a paved road connecting distant towns. Freeways are multi-lane highways that use on- and off-ramps, rather than intersections, in order to limit the number of entrance and exit points along the route, hence keeping traffic along the freeway fairly steady.

When was the **first freeway** built in the United States?

The first freeway (lacking tolls and having limited access) in the United States was the Arroyo Seco Freeway, connecting Pasadena and downtown Los Angeles. It opened in 1940 and is now the Pasadena Freeway, Highway 110.

What are **interstate highways**?

President Dwight D. Eisenhower signed the Federal-Aid Highway Act of 1956, which established the system of interstate highways in the United

States. Interstate highways are federally funded freeways that allow the rapid transportation of people, goods, and the military across the country.

Why was **Eisenhower a fan** of interstate highways?

In 1919, the young Dwight D. Eisenhower took part in a cross-country military trip from Washington, D.C., to San Francisco. But, due to the state of the highways at that time, the trip took 62 days—far too long to defend the country should the need arise. This experience made Eisenhower realize the need for a faster, more efficient mode of transportation across the country. Because of President Eisenhower's support for the Interstate Highway System, it is now officially known as the Dwight D. Eisenhower System of Interstate and Defense Highways.

Why does **Hawaii** have interstate highways?

Since any freeway funded under the Federal-Aid Highway Act of 1956 is known as an interstate highway, whether it crosses state boundaries or not, Hawaii can have interstate highways. Though they cross no state borders, Hawaii has three interstate highways, H1, H2, and H3.

When was the **last interstate highway** built?

The construction of new interstate highways came to an end in 1993 with the opening of Interstate 105, the Century Freeway, in Los Angeles, 37 years after construction began on the system. The Century Freeway is an inter-city route connecting the coastal community of El Segundo to Interstates 405, 110, 710, and finally 605 in Norwalk.

How many **miles of paved road** are there in the United States?

The United States has more paved road than any other country in the world, with a grand total of 2,335,000 miles.

111

British and French diggers shaking hands upon completing the Channel Tunnel. (Reuters/Archive Photos)

Did Hitler create the **Autobahn**?

Though the first modern freeway system in Germany was begun in 1913, Adolf Hitler did create the Autobahn during the Third Reich, from 1933 to 1945. The Autobahn is a freeway system that includes 6,800 miles of road across Germany. Though it is widely believed that there are no speed limits on the Autobahn, there are a few segments with marked speed limits.

What is the **longest bridge** in the world?

Completed in 1969, the Lake Pontchartrain Causeway that connects New Orleans with Mandeville, Louisiana, is 24 miles long—the longest bridge in the world.

What is the **Chunnel**?

The Channel Tunnel (or Chunnel) is a railroad tunnel under the Strait of Dover in the English Channel. The Chunnel runs for 31 miles between Folkestone (near Dover) in the United Kingdom and Sangatte (near Calais) in France. Opened in 1994, the Chunnel connects England with the rest of continental Europe.

What is the **most common street name** in the United States?

It's not Main Street. Park is actually the country's most common street name, followed by Washington, Maple, Oak, and Lincoln.

Where is the **longest Main Street** in the United States?

Main Street in Island Park, Idaho, is 33 miles long, making it the longest in the United States.

When was the **first automobile** built?

Though it had only three wheels, the world's first gasoline-powered automobile was built by Karl Benz in 1885. Henry Ford built his in 1893.

Which city has the most **taxis**?

Congested Mexico City is home to more than 60,000 taxis among its 3.5 million automobiles.

Where was the **first self-service gas station**?

In the automobile city of Los Angeles, George Urich opened the first self-service station in 1947.

Who invented the **traffic signal**?

The red, yellow, and green traffic signal that we are familiar with today was originally invented by Garrett Morgan in 1923. Morgan, who was also the inventor of the gas mask, received numerous awards for his invention.

What is the world's longest subway system?

The world's first subway system is also the world's longest. In operation since 1863 (when it used steam-powered trains), the London Underground is now 244 miles long and uses electricity to power the trains.

Who invented the **first train**?

In 1825, British engineer George Stephenson invented the first train, which was powered by steam. Stephenson's train was introduced to North America in the 1830s and was used until the 1940s when diesel-electric locomotives, which didn't run on expensive coal, replaced steam locomotives.

Sea Transport

What is the world's **busiest seaport**?

The world's busiest seaport is Rotterdam, Netherlands, which handles over 325 million tons of cargo each year. The Port of South Louisiana in New Orleans is the world's second-busiest port, with over 200 million tons of cargo annually. The next three busiest ports are Singapore; Kobe, Japan; and Shanghai, China.

What does a **lock on a canal** do?

Many canals connect two bodies of water that lie at different elevations. Locks are used to gradually move the ships from one elevation to another. Once a ship enters a lock, doors close in front of and behind it. Water is then added or drained from the area to raise or lower the ship

The Panama Canal. (Archive Photos)

to a different elevation. Then the doors in front of the ship open and the ship sails down the canal to the next lock or to the open sea.

In which **direction** do ships sail **through the Panama Canal**?

Though you would expect them to travel east from the Pacific to the Atlantic Ocean when sailing through the Panama Canal, ships actually travel northwest. Since the Isthmus of Panama lies parallel to the equator, the canal does not lie east-west but rather northwest-southeast.

Why was the **Erie Canal** built?

The 363-mile Erie Canal connects the Hudson River to Lake Erie. Opened in 1825, the canal created a new, shorter route from the northern interior of the United States to the Atlantic Ocean. Prior to the opening of the Erie Canal, goods traveled down the Mississippi River and out to the Atlantic. Since New York City lies on the Hudson River, the canal was responsible for the growth of the city as a major port, helping it to become the largest city in the United States. Once the St. Lawrence Seaway was built in 1959, the Erie Canal became rarely used, since most transportation soon traveled along the Seaway.

What is the **St. Lawrence Seaway**?

Completed in 1959, the 183-mile-long St. Lawrence Seaway was built by deepening and widening the St. Lawrence River between Montreal, Canada, and Lake Ontario so that large ships could traverse it. The Seaway consists of a series of locks that allow ships to travel from the Atlantic Ocean to the Great Lakes, and ultimately on to Chicago. A limiting factor of the Seaway is that ships can only use the Seaway between May and November, as it is blocked by ice in the winter.

POLITICAL GEOGRAPHY

How does **geography influence politics**?

Geography is a key component in many political decisions and actions. The borders of countries, location of natural resources, access to ports, and the designation of voting districts are a few of the many geographical factors that affect politics.

What is the difference between a **country and a nation**?

Many people use the terms "country" and "nation" interchangeably. But not all nations are countries, nor are all countries nations. A country is the equivalent of a State, and is a political entity. A nation is a group of people with a common heritage and culture. Some nations have a State and are thus called a nation-states. Nation-states include France, Germany, Japan, China, and the United States. Some nations have no State, such as the Kurds and Palestinians. Some States have multiple nations such as Belgium, which is composed of two nations, the Flemings and Walloons.

What is the difference between a **State and a state**?

A State, with a capital "S," is equivalent to a country. A state, with a lower case "s," is a division of a country, like the states that make up the U.S.

Who owns the world's oceans?

The battle over control of the world's oceans has increased over the past few decades due to the discovery of vast mineral and fuel resources located under the sea. In 1958, the United Nations held the first Conference on the Law of the Sea. This conference established territorial seas, measuring 12 nautical miles from the shore of coastal nations, that are under the full control of that country. Additionally, countries have mineral, fuel, and fishing rights in an Exclusive Economic Zone (EEZ) that spans 200 nautical miles from shore. Problems arise when two countries' zones overlap. Median lines between countries have been drawn in most cases, but there are still many areas of disagreement.

Do all countries **have states**?

While most countries are divided into states, provinces, or departments, there are many that have no political divisions. Large countries without political divisions include Mali, Kazakhstan, Saudi Arabia, and Algeria.

How does a **choke point** "choke" a body of water?

A choke point is a narrow waterway between two larger bodies of water that can be easily closed or blocked to control water transportation routes. Though historically the Strait of Gibraltar (connecting the Mediterranean Sea and Atlantic Ocean between Africa and Spain) has been one of the world's most important choke points, the Strait of Hormuz gained significant attention during the Persian Gulf War of 1991. The Strait of Hormuz, bounded by the United Arab Emirates and Iran, connects the Persian Gulf to the Arabian Sea and, thus, to the Indian Ocean. It was feared that if Iraq controlled the Strait, then most of the oil from the region could not be shipped out.

Who controls the **world's oil supply**?

The Organization of Petroleum Exporting Countries (OPEC) coordinates most of the world's oil production. The members of OPEC meet to coordinate oil policies and prices. Twelve of the world's leading oil-producing countries are members: Algeria, Gabon, Indonesia, Iran, Iraq, Kuwait, Libya, Nigeria, Qatar, Saudi Arabia, United Arab Emirates, and Venezuela. Though Russia, the United States, and Mexico are also leading petroleum producers, the three countries are not members of OPEC.

THE UNITED NATIONS AND NATO

THE UNITED NATIONS

How does the **United Nations preserve peace**?

Established in 1945 at the end of World War II, the United Nations was created for the purpose of maintaining world peace. Its members pledge to work together to solve disputes. The U.N. also oversees many agencies that promote health, welfare, and cooperation around the world.

How many countries are **members of the United Nations**?

There are currently 185 members in the U.N. The most recent addition is the island country of Palau, which joined in 1994.

Which countries are **not members** of the United Nations?

While almost every country in the world is a member of the U.N., there is a short list of countries that are not members: Kiribati, Nauru, Switzerland, Taiwan, Tonga, Tuvalu, and the Vatican (which is not technically considered a country).

Why do some countries **choose not to join the United Nations**?

A country must be willing to give up some of its self-rule for the greater good, which often means the good of the larger, western countries. Switzerland's professed neutrality keeps it from joining.

How did the **League of Nations** fail?

The League of Nations, which was created in 1920 and replaced by the United Nations in 1945, failed in its mission to prevent the Second World War. Even though the League was essentially the creation of President Woodrow Wilson, the isolationist United States never became a member.

NATO AND THE COLD WAR

Which countries are **members of NATO**?

In 1998, the 16 members of NATO included Belgium, Canada, Denmark, France, Germany, Greece, Iceland, Italy, Luxembourg, Netherlands, Norway, Portugal, Spain, Turkey, the United Kingdom, and the United States. The three former Communist countries that are expected to join NATO in 1999 are Poland, Hungary, and the Czech Republic.

How did the **Communists** respond to the creation of NATO?

In 1955, seven Communist countries created the Warsaw Pact to protect against NATO aggression. The Warsaw Pact originally consisted of Albania, Bulgaria, Czechoslovakia, Hungary, Poland, Romania, and the Soviet Union. The Warsaw Pact disbanded in 1991 with the breakup of the U.S.S.R. and the changes in Eastern Europe.

What is the **purpose of NATO** now that the Soviet Union is gone?

The North Atlantic Treaty Organization (NATO) was founded in 1949 as an alliance of European and North American non-Communist countries committed to preventing and protecting against Communist threats.

Inside the United Nations. (Archive Photos)

Now that the threat of Soviet aggression has been reduced, NATO sees its role as a policing agency for Europe.

What did the **domino effect** have to do with U.S. involvement in the Vietnam War?

American military strategists believed that if one country became Communist, it would begin a never-ending succession of countries converting to Communism (thus the domino metaphor). Policy makers believed that the U.S. had to do everything possible to keep every country from falling to Communism. This included sending American troops to Vietnam. Though Vietnam fell to Communism, the theory of the domino effect was proven incorrect because neighboring countries did not fall to Communism as predicted.

Who won the **Cold War**?

Beginning in the years following World War II, the Cold War was the battle for power between the United States and the Soviet Union. The

A Soviet military parade on Red Square. (Reuters/Archive Photos)

Cold War spurred the development and stockpiling of nuclear weapons, intense militarization, division between eastern- and western-bloc nations, and the race to outer space. The stalemate of the Cold War effectively ended when the Berlin Wall fell in 1989. Two years later, Communism in the Soviet Union collapsed. Neither the United States nor the Soviet Union won this "war," in which no battles were fought.

THE WORLD TODAY

What are the world's **newest countries**?

In the 1990s, over *two dozen* new countries appeared on the map. These included 15 new countries that were created when the U.S.S.R. broke up in 1991—Armenia, Azerbaijan, Belarus, Estonia, Georgia, Kazakhstan, Kyrgyzstan, Latvia, Lithuania, Moldova, Russia, Tajikistan, Turkmenistan, Ukraine, and Uzbekistan.

The dissolution of Yugoslavia also created several new countries in 1991 and 1992—Bosnia and Herzegovina, Croatia, Macedonia, Serbia and Montenegro, and Slovenia.

In 1993, Eritrea became independent of Ethiopia, and Czechoslovakia dissolved into the Czech Republic and Slovakia. That same year, the Pacific island countries of the Marshall Islands, Micronesia, and Palau all became independent. In 1990, Namibia split from South Africa to become its own country, while East and West Germany combined to become Germany.

How many countries does the United States recognize?

The State Department is the official United States government agency that recognizes independent countries. It maintains an updated list of the official independent States of the world. As of this writing, there are 190 States listed. Taiwan and the Vatican, though commonly considered countries, are not included on this list.

Why isn't **Taiwan recognized** by the U.S. government?

Though Taiwan would like to be considered an independent country, China claims Taiwan as part of its territory. Thus, Taiwan does not meet the criteria for independent status because it does not have its own territory.

Why are some **borders curvy while others are straight**?

There are two primary types of boundaries—geometric and natural. Geometric boundaries are straight and follow lines of latitude, longitude, or a certain compass direction between points. Geometric boundaries were established to divide territories before settlers entered areas. Most of the states in the Western United States have at least a portion of their borders formed by geometric boundaries (especially rectangular-shaped Colorado and Wyoming). Natural boundaries are usually curvy because they follow the crests of mountains or the center of rivers. Natural boundaries are very common in places like Europe, where the region was populated before the countries were created.

Why is the border between the **Yemen and Saudi Arabia** dashed on maps?

Some borders between countries have yet to be determined. In the sandy desert between Saudi Arabia and Yemen, the border between the

Why do third world countries no longer exist?

The term "third world" was part of the classification of countries during the Cold War. This classification designated those countries aligned with the United States as "first world," those countries aligned with the Soviet Union as "second world," and those countries that were nonaligned as "third world." Over time, the term "third world" came to mean a poorer or less-developed country. With the dissolution of the Soviet Union and abandonment of Communism by Eastern Europe, the classification no longer exists. The preferred terms are now "developed" countries and "less-developed" (or "developing") countries.

two countries is disputed and has not yet been defined. So the border between Saudi Arabia and Yemen appears as a dashed line on many maps.

Which countries are surrounded entirely by **landlocked countries**?

The two countries surrounded entirely by landlocked countries are Uzbekistan and Liechtenstein. Uzbekistan is surrounded by the landlocked countries of Kazakhstan, Kyrgyzstan, Tajikistan, Afghanistan and Turkmenistan. Liechtenstein is bordered by Switzerland and Austria, neither of which have access to the ocean.

How is **gerrymandering like a salamander**?

In 1812, Massachusetts Governor Ellbridge Gerry signed a law that established an oddly shaped Congressional district. It was redrawn by political cartoonists into a salamander-type creature and thus the term gerrymander was born. Gerrymandering is the process of establishing oddly shaped congressional districts in order to include voters from dispersed areas. Gerrymandered districts can be helpful or detrimental to minority groups, depending on who draws the borders. The U.S. courts have found gerrymandering to be a legal method of establishing congressional boundaries.

What is the **best shape** for a country?

Though countries come in various shapes for various reasons, the best shape for a country is compact. A compact country, such as Germany and France, is easier to govern than those that are fragmented (such as Indonesia) or elongated (such as Chile). Compact countries are easier to govern because transportation, communication, and internal security are easier to maintain. Also, compact countries have shorter borders to protect. Elongated and fragmented countries are more easily divided and conquered.

COLONIES AND EXPANSIONISM

Why did the sun never set on the **British Empire**?

In the early 20th century, the United Kingdom included colonies from North and South America (Canada, British Guiana, and Bermuda), Africa (Egypt, South Africa, and Nigeria), Asia (India and Burma), and Oceania (Australia and New Zealand). Because the British Empire spanned the globe, there was always at least one portion of the Empire in daylight.

Why would countries want **colonies**?

Colonies are a source of raw materials, new land, wider trading opportunities, and militaristic expansion for the mother country. Colonies were established around the world from the 16th century through 19th century by powerful western nations. After World War II, the concept of colonization was widely attacked as an exploitive policy. Though most colonies were granted independence, many countries still control colonies around the world.

Adolf Hitler watches the Nazi empire expand in the early days of World War II. (The National Archives/Corbis)

What is the **largest colony** in the world?

Puerto Rico, a colony of the United States with approximately 3.6 million residents, is currently the largest colony in the world. In 1999, Puerto Ricans may have the opportunity to vote whether they would like to become a state of the U.S., an independent country, or remain a colony. If Puerto Rico becomes the 51st state or becomes its own country, then French Polynesia would become the world's largest colony. French Polynesia is a group of islands located in the South Pacific Ocean that have about 200,000 residents.

How did the **Nazis use geopolitics**?

During the Nazi era in Germany, 1933–1945, the "science" of geopolitics was utilized to support Germany's concept of Lebensraum, or living space. The Nazi concept of Lebensraum was based on the idea that there was a racial heirarchy that allowed "superior" races to conquer "inferior" races. Adolf Hitler used this warped sense of geopolitics to invade Czechoslovakia, Poland, and the Soviet Union. For example, Germany

127

claimed that the ethnic Germans living in the Sudentenland of Czechslovakia should be included within the German fatherland.

How did **irredentism** help start World War II?

Irredentism is a term used to describe a situation in which a minority group in one country shares the culture and heritage of another country. The minority group may attempt to have their region annexed into the mother country or may be happy in the country they are in. Adolf Hitler used irredentism as an excuse to invade and conquer Czechoslovakia in 1938. He claimed that the Germans in the Sudetenland, a part of Czechoslovakia, were being treated unfairly and thus this area should be annexed to Germany. Though Germany's annexation of Czechoslovakia did not start World War II, it was the Nazis' first direct aggressive step toward conquering Europe.

THE WORLD ECONOMY

What is the difference between **GNP and GDP**?

GDP, or gross domestic product, is the value of all goods and services produced in a country in a year. GNP, or gross national product, is the total value of GDP plus all income from investments around the world. Per capita GDP is usually compared between countries.

Which countries give the highest proportion of their GNP in **foreign aid** to other countries?

Norway, Denmark, Sweden, Netherlands, and Finland each give about one percent of their GNP to other countries as foreign aid.

Which country is the world's leading **importer and exporter**?

The United States, leading both categories, is responsible for 12 percent of the world's exports and 15 percent of the world's imports.

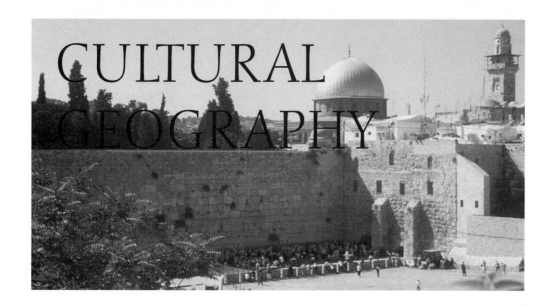

CULTURAL GEOGRAPHY

POPULATION

Of all the humans who have ever lived, what proportion of them are **alive today**?

Only a small percentage, anywhere from 5 to 10 percent, of the humans who have ever lived are alive today. Since humans have existed for approximately 100,000 years, the total number that have ever lived is probably between 60 billion and 120 billion.

How many people **live on the Earth**?

There are approximately 6 billion people on the Earth. The planet's population is expected to grow to 8 billion by the year 2025 and 9.3 billion by 2050.

What has the **world's population** been over time?

Year	Population
0	200 million
1000	275 million
1500	450 million

Year	Population
1750	700 million
1850	1.2 billion
1900	1.6 billion
1950	2.6 billion
1960	3 billion
1975	4 billion
1985	4.85 billion
1990	5.3 billion
1999	6 billion

How fast is the world's **population growing**?

In 1998 the annual natural increase of the world's population was approximately 1.3 percent (79 million people). That represents 133 million births and 54 million deaths annually, which averages out to be 4.2 births every second and 1.7 deaths every second.

Which country has the world's highest **population density**?

Monaco, with an area of just 0.75 square miles, has the world's highest population density, with over 42,500 people per square mile. The country with the highest population density that is not a mini-state (like Monaco) is Bangladesh, with a population density of approximately 2,200 people per square mile.

What is a **census**?

A census is an enumeration, or counting, of a population. The information from a census is used to help governments determine where to provide services, based on the demographics of the population. Information about age, gender, number of children, race, languages spoken, education, commuting distance, salary, and other demographic variables is common in a census. This information is compiled and provided to government agencies and is usually accessible to the general public.

In the United States and most other developed nations, a census takes place once every decade. The Constitution of the United States requires

Monaco, with an area of just 0.75 square miles, has the world's highest population density, with over 42,500 people per square mile. (Jonathan Blair/Corbis)

a census to be taken every 10 years, in order to create districts and determine the number of members of Congress each state is able to send to the House of Representatives.

What was the **baby boom**?

Due to post–World War II prosperity, there was a boom in U.S. births between 1946 and 1964, now referred to as the "baby boom." During this time, approximately 77 million babies were born in the United States, a very large number compared to that of other time spans. As the baby-boomers approach retirement age, health and welfare services for the elderly will become high priority as the country prepares for a higher proportion of older people in its population than ever before.

Are the an **equal number of boy and girl** babies born?

Though scientists aren't sure why it occurs, there is an average of 105 boys born for every 100 girls.

LANGUAGE AND RELIGION

What are the most **commonly spoken languages** of the world?

The world's most commonly spoken language, Mandarin, is spoken by over one billion people in China, Taiwan, and Singapore. English (nearly 500 million), Hindi (more than 450 million), Spanish (approximately 400 million), and Russian (over 250 million) are the other languages most commonly spoken around the world.

What is the difference between a **lingua franca and a pidgin**?

A lingua franca is a language used between people who do not have a common language. English is often used as a lingua franca in international business transactions. A pidgin is a language that has a small vocabulary and is a combination and distortion of two or more languages. For example, pidgin English, a combination between English and indigenous languages, is used in Papua New Guinea between English-speaking and indigenous people. Most pidgins are lingua francas but not all lingua francas are pidgins.

Which **religions** have the **most adherents**?

There are over 1.1 billion Muslims in the world, making Islam the world's most popular religion. Roman Catholics (over 950 million), Hindus (nearly 800 million), Protestants (over 400 million), and Buddhists (approximately 325 million) belong to the other most popular religions. Nearly 2.7 billion people are atheists, agnostic, or indifferent to religion.

How many religions have holy sites in **Jerusalem**?

Judaism, Islam, and Christianity all regard Jerusalem as a very holy city. The Western Wall, the remaining wall of the Second Temple, is the holiest site in Judaism. Islam's third-holiest site is the Dome of the Rock and the mosque, both located in Jerusalem. The Church of the Holy Sepulcher is a holy Christian site.

What were Thomas Malthus' ideas on population growth?

In 1798, English clergyman Thomas Malthus wrote "An Essay on the Principle of Population," in which he described the problems of population growth. Malthus argued that the world's population grows faster than the food supply, but there are such checks as war, famine, disease, and disaster that limit the population.

Why do **mosques have domes**?

The onion-shaped domes of Islamic mosques and other religious buildings of eastern religions were an architectural style borrowed from the Byzantine Empire. One of the world's most famous onion-domed buildings, St. Basil's Cathedral in Moscow's Red Square, was built in the mid-16th century.

DEALING WITH HAZARDS

Is **childbirth** still a significant cause of death for women?

From the beginning of time until the mid-20th century, a leading cause of death for young women was complications during childbirth. Now, in developed countries, there is almost no risk of death during pregnancy and labor.

How did the **Black Plague** affect the world's population?

Spread by fleas, the bubonic plague, also called the Black Plague, raged through Europe, Asia, and North Africa between 1346 and 1350.

135

The Western Wall and Dome of the Rock in Jerusalem. (Archive Photos)

136

Is there enough food to feed the world?

Though there is enough food produced in the world to feed everyone, logistical and political problems make its distribution inefficient. At the current rate of world population growth, we may soon have to change our eating habits and eat more grain and less meat. There is a limited amount of grain that the Earth can produce. Currently, much of this grain is consumed by cattle, rather than humans. If humans were to eat the grain instead of eating the cattle, the calories from the grain would be twenty times more efficent than those from beef.

Though cities attempted to curb the spread of this highly infectious disease by quarantining cities, the fleas easily spread from city to city. Estimates of those killed reach into the tens of millions. In Europe and Asia, more than half of the population died of the Black Plague during those four years. Many more died of starvation in the famine that followed because of the staggering depletion of the work force.

How widespread was the **influenza pandemic of 1918**?

In 1918, a deadly flu spread quickly around the world. Within just two years, this influenza pandemic had sickened over a billion people and killed more than 21 million. Half a million people died in the United States alone.

What **revolution** attempted to stop **world hunger**?

Begun in the 1960s, the "Green Revolution" was an attempt by developed countries and such international organizations as the United Nations to transfer agricultural technology to less-developed countries. While the Green Revolution increased agricultural yields, it modified

137

What is a refugee?

A refugee is a person who leaves his or her home country for fear of persecution. There are approximately 15 to 20 million refugees in the world today. Most refugees come from developing countries where society is in flux or even chaotic. Refugees usually flee to the closest stable country, so different countries see great variation in the number of refugees based on the political climate of their neighbors. Thus, developed countries consistently have a large number of refugees arriving. Refugees are problematic, just as any mass immigration is.

the ecology of traditional agricultural systems (such as through the use of chemical fertilizers) and has yet to cure world hunger.

How does **medical geography** help control the spread of diseases?

Medical geographers and epidemiologists (scientists who study diseases and epidemics) use mapping to monitor the spread of diseases and locate the source of a disease. For example, by mapping a group of inordinately high numbers of cancer patients in a city, we may find that all live close to a factory that has been releasing toxins into the groundwater. By identifying the source and spread of a disease, the disease can often be combated.

CULTURE AROUND THE WORLD

What are **nomads**?

Nomads are tribes that migrate in a seasonal circuit over a large region. Though nomadic people often build temporary homes, they consider migratory life within their tribe to be home. Nomadic tribes are located in marginal areas around the world, from the Sahara Desert to northern Siberia. The nomadic way of life is threatened because of general cultural prejudice against unsettled peoples.

What is a **Gypsy**?

Gypsies are nomadic tribes that travel throughout Europe. Though they once were thought to have originated in Egypt (hence the European name for them), linguistic studies have placed their origin in India. These traveling tribes have been subject to centuries of persecution, including "Gypsy hunts" and extermination at the Auschwitz Death Camp.

What is **brain drain**?

When highly educated or highly skilled individuals leave their home countries to go to countries where opportunities are better, the home country experiences "brain drain." This occurs especially in Asian countries, as highly educated Asians move to the United States, Canada, and Australia for higher-paying jobs.

How do people cope with **continual light or darkness** in high latitudes?

Murmansk, Russia, is the largest city north of the Arctic Circle. The city receives no sunlight for several months out of the year, making it one of the most psychologically extreme environments on the planet. Residents of the city (about 470,000) walk along artificially lit streets that give the appearance of sunlight, undergo artificial sun treatments (much like tanning booths), and often suffer from the condition known as Polar Night Stress. Polar Night Stress symptoms include fatigue, depression, vision problems, and susceptability to colds and flus.

What is a **long lot**?

Long lots are long and narrow pieces of property. This type of division of land is common in Europe and places in North America that were initial-

139

Why do some people eat dirt?

The practice of eating dirt, called geophagy, is most commonly practiced by pregnant and lactating women. Since women's bodies require additional nutrients during pregnancy and lactation, the body craves clays and dirts that carry these additional minerals. This practice is most common in Africa, but the practice spread to the United States with the forced migration of Africans during slavery. Geophagy is now also practiced in the southern United States, but the practice has become a cultural rather than physiological action.

ly settled by the French (such as Québec and Louisiana). Each lot has a narrow access to a stream or road but is several hundreds of feet deep.

Why do some cultures **kill infants**?

Infanticide is the practice of killing an infant. For centuries, various cultures around the globe used infanticide as a form of population control, most commonly because their limited food supply could only feed a certain number of humans. Because of cultural biases, female infants were more often victims of infanticide. The practice of infanticide still occurs today.

Why don't Americans eat **horse meat**?

Most religions and cultural groups have some kinds of food taboos. Foods may be avoided entirely or may be avoided on certain days or during certain festivals. Religious food taboos include the avoidance of pork by Muslims and Jews and the avoidance of beef by Hindus. Cultural food taboos also play an important role. For example, Americans don't eat horse meat because it is a cultural food taboo, despite the fact that horse meat is a nutritional and edible type of food.

Tokyo, Japan: site of just one of the world's 23,000 McDonald's restaurants. (Archive Photos)

How many **McDonald's restaurants** are there in the world?

Since the opening of its first restaurant in 1948 in San Bernardino, California, McDonald's has grown into a world-wide restaurant chain. There are now over 23,000 McDonald's restaurants in 109 countries.

Why does a **first-born son** get everything?

Primogeniture is the system of inheritance in which all inheritable land and property is passed on to the first-born son. A common worldwide tradition, primogeniture enabled a family's possessions and status to remain intact as they were passed from generation to generation. This practice of the entire inheritances benefiting only the first-born son resulted in subsequent sons needing to find alternative livelihoods.

Can a woman have **multiple husbands**?

While some cultures allow men to have multiple wives, there are other cultures that allow women to have multiple husbands. This practice,

141

known as polyandry, is presently observed by only two cultures, the Tibetans and the Nair people of southwestern India. Polygyny, when men have multiple wives, remains legal in Islamic and many African countries. The collective name for multiple spouses, both polyandry and polygyny, is polygamy.

When did people start eating with **forks and spoons**?

Though introduced in the 15th century, forks and spoons did not come into common use in Europe until the 17th century. Prior to that time, people ate with their hands and a knife.

What is the most common **last name** in the world?

The world's most common surname is Chang.

TIME, CALENDARS, AND SEASONS

Why did early humans have **no need for hours, days, weeks, or months**?

Because early humans were hunters and gatherers, they had little need to know exact time. What *was* essential was an understanding of the seasonal migration of animals and varieties of plant life.

TIME

What time is it at **12:00 a.m.**?

12:00 a.m. is midnight. In the middle of the night, 11:59 p.m. is followed by 12:00 a.m., not 12:00 p.m., which is noon.

What do **"a.m." and "p.m."** mean?

The distinctions "a.m." and "p.m." are abbreviations for "ante meridiem" and "post meridiem," which mean before midday and after midday, respectively.

143

What's the difference between Greenwich Mean Time, Universal Coordinated Time, and Zulu Time?

Greenwich Mean Time (GMT), Universal Coordinated Time (UTC), and Zulu Time are three different names for the same time zone. This time zone is situated at the Prime Meridian, zero degrees longitude, and runs through the Royal Observatory at Greenwich, a section of London, England. It is the Prime Meridian from which other longitudes are determined, east and west.

How does **military time** work?

Often known as military time, 24-hour time is used by many countries. It begins at midnight with 0000 and the day ends with 2359. The first two digits represent the hour and the last two digits represent the minutes. Since there are 24 hours in a day, each hour is numbered 00 through 23. For instance, 0100 is 1:00 a.m., 1200 is noon, 1300 is 1:00 p.m., and 2043 is 8:43 p.m.

How **long is a day**?

A day is the time it takes the Earth to make one rotation, which is 23 hours, 56 minutes, and 4.2 seconds. We round this to an even 24 hours for convenience.

Does the Earth always **rotate and revolve around the sun** at the same speed?

No, the Earth's revolution around the sun and rotation on its axis are not perfect. Its daily rotation may vary by approximately four to five milliseconds, and is slowing down at a rate of one millisecond each century, due to tidal friction. Additionally, the axis wobbles slightly, and the length of revolution around the sun varies by a few milliseconds.

How can I find the **exact time**?

Soon after use of the telegraph began, correct time signals were broadcast across their wires. Today, you can often call a special phone number to obtain the correct time or tune in to the U.S. National Institute of Technology and Standards short wave radio station at 2.5, 5, 10, 15, and 20 megahertz.

TIME ZONES

When were **time zones established** in the United States?

In 1878, Sir Sanford Fleming proposed dividing the world into 24 time zones, each spaced 15 degrees of longitude apart. The contiguous United States was covered by four time zones. By 1895, most states had begun to institute the standard time zones of Eastern, Central, Mountain, and Pacific on their own. But it wasn't until 1918 that Congress passed the Standard Time Act, establishing official time zones in the United States.

How did **trains** help establish time zones?

Before trains, many cities and regions had their own local time, which was set based on the sun at their location. The great variation of local times made train schedules confusing. In November, 1883, railroad companies across the U.S. and Canada began to use standard time zones— years before they came into general use across the United States.

How many time zones does the United States have?

The United States spans nine time zones: Eastern, Central, Mountain, Pacific, Alaska, Hawaii-Aleutian, Samoa, Wake Island, and Guam.

A 24-hour clock at the C.I.A. (Roger Ressmeyer/Corbis)

Which **states are split** into multiple time zones?

Florida, Indiana, Kentucky, and Tennessee are split into Eastern and Central time. Kansas, Nebraska, North Dakota, South Dakota, and Texas are split between Central and Mountain time. Idaho and Oregon are split between Mountain and Pacific time.

How many time zones does **China** have?

Since China is such a large country, it should span five time zones, but the entire country uses one time—eight hours ahead of UTC.

Why did the International Date Line recently move?

Prior to 1995, the country of Kiribati, consisting of 21 inhabited islands, straddled the International Date Line, thus dividing the country into two days. In 1995 Kiribati decided to shift their portion of the International Date Line far to the east so that the entire country could be on the same side of the Date Line.

What time is it at the **North and South Poles**?

Because time zones get narrower the farther you get from the equator, time zones would be very thin near the North and South Poles. To simplify things, researchers living in Antarctica use Coordinated Universal Time (UTC), which is the time at Greenwich, England.

What happens when I cross the **International Date Line**?

If you fly, sail, or swim across the International Date Line from east to west, such as from the United States to Japan, you add a day (Sunday becomes Monday). When you travel from west to east, such as from Japan to the United States, you subtract a day (Sunday becomes Saturday).

How fast do you have to travel west to **arrive earlier** then when you left?

Normally, when flying between London and New York, the trip takes seven hours. Thus, with the five-hour difference in time between the two cities, you arrive two hours "later" than when you left London. If you were to fly on the Concorde, which travels at Mach 2 (1,300 miles per hour—two times the speed of sound), the trip between London and New York takes only three hours. Thus, with the five-hour difference in time zones, you will arrive two hours "earlier" than when you left!

147

Why is **Russia always one hour ahead**?

In an effort to take advantage of the limited amount of light available in winter months, each of Russia's time zones are one hour ahead of the standard time for those zones. Russia also follows Daylight Saving Time and adds an additional hour during spring and summer months.

DAYLIGHT SAVING TIME

Why do we have **Daylight Saving Time**?

By moving our clocks forward one hour between spring and fall, we more effectively utilize the light of the sun to keep homes and businesses lit, saving electricity.

When does **Daylight Saving Time start and end** in the U.S.?

Daylight Saving Time begins at 2:00 a.m. on the first Sunday in April and ends at 2:00 a.m. on the last Sunday in October. During this time the clock is advanced one hour ahead.

1998	2 a.m. April 5 to 2 a.m. Oct. 25
1999	2 a.m. April 4 to 2 a.m. Oct. 31
2000	2 a.m. April 2 to 2 a.m. Oct. 29
2001	2 a.m. April 1 to 2 a.m. Oct. 28
2002	2 a.m. April 7 to 2 a.m. Oct. 27
2003	2 a.m. April 6 to 2 a.m. Oct. 26

When did Daylight Saving Time move from the end to the **beginning of April** in the U.S.?

The shift of the beginning of Daylight Saving Time from the last Sunday in April to the first Sunday in April took place in 1987, when the Uniform Time Act was amended.

When was Daylight Saving Time instituted?

Though Benjamin Franklin suggested the concept of Daylight Saving Time in 1784, it was not implemented in the United States until World War I. Between World Wars I and II, states and communities were allowed to choose whether or not to observe the change. During World War II, Franklin Roosevelt again implemented Daylight Saving Time. Finally, in 1966, Congress passed the Uniform Time Act, which standardized the length of the Daylight Saving Time period. But states and territories can choose not to observe Daylight Saving Time. Arizona, Hawaii, parts of Indiana, Puerto Rico, and some island territories have chosen not to observe Daylight Saving Time.

When do countries in the **Southern Hemisphere** observe Daylight Saving Time?

Because Daylight Saving Time is an effort to save daylight during the summer months, Daylight Saving Time in the Southern Hemisphere occurs from October through March.

KEEPING TIME

What is a **sundial**?

A sundial is an instrument that uses the sun to measure time. A sundial consists of an angled marker, called a gnomon, that casts a shadow on a plate, called a dial plane. On the dial plane there are marks indicating the hours of the day. During the day, as the sun moves across the sky, the shadow from the gnomon moves across the dial plane, indicating the hour. Sundials were used to measure time before clocks and watches were invented.

Sundials use the sun to measure time. (Archive Photos)

What is a **water clock**?

Water clocks were the first clocks that didn't rely upon sunlight (as with sundials) to tell time. They operated by dripping water from containers at measured intervals. There were two key types of water clocks: those that measured time by the amount of water remaining in the clock and those that measured time by how much water dripped from the clock and filled a measuring device.

When was the **first watch** made?

In the early 16th century, German locksmith Peter Kenlein began to produce portable clocks called Nürnberg eggs. Advances in later centuries led to timepieces that could be worn on the wrist.

What is an **atomic clock**?

An atomic clock uses measurements of energy released from atoms to precisely measure time. The current model of the atomic clock, created in 1957 by Norman Ramsey, uses measurements from cesium atoms. Atomic clocks are used by NASA, physicists, astronomers, and other scientists who need extremely precise time.

CALENDARS

What do B.C. and A.D. stand for?

In our modern calendar, the year 0 represents the year of the birth of Jesus Christ. Years before his birth are known as B.C., or "before Christ." Years since his birth are known as A.D. or "Anno Domini," the "Year of Our Lord."

What do B.C.E. and C.E. stand for?

To secularize the calendar, the terms B.C.E. and C.E. have come into use to replace B.C. and A.D. respectively. B.C.E. means "before common era" and C.E. stands for "common era."

What is the problem with a **calendar based on the cycles of the moon**?

The time between two new moons is 29 and one half days. After 12 lunar months, a calendar based on the cycles of the moon falls short of a solar year—and thus the cycle of seasons—by 11.25 days. To compensate, the Hebrew Calendar, which is based on the moon cycle, has a regulated 19-year cycle in which an extra month is added every two or three years.

How did **Julius Caesar** fix the calendar?

For years, the Romans had been using a calendar based on lunar cycles. Since each lunar month is 29.5 days, 12 months only adds up to 354

What was the longest year in history?

The year 46 B.C.E. was decreed by the Roman Emperor Julius Caesar to be 445 days long in order to correct the calendar, which was 80 days off, based on the seasons.

days. But seasons do not follow a lunar cycle, they follow a solar one. A solar year lasts 365 days, 5 hours, 49 minutes. Julius Caesar implemented a solar calendar so that the seasons would occur at the same times every year. Additionally, Caesar made each year 365 days long, with every fourth year being 366 days long (a "leap" year). Unfortunately, each calendar year was was still 11 minutes longer than a solar year, a problem that Caesar did not feel was a big concern at the time.

When was **January 1st** chosen as the beginning of the year?

In Caesar's calendar modifications of 46 B.C.E., he decreed that the year would begin on January 1st instead of March 25th, as it had in the past. At that time Caesar also designated number of days in each month, unchanged to the present day.

Why were **10 days lost** from the year in 1582?

In 46 B.C.E., Julius Caesar implemented the Julian calendar, which was 11 minutes longer than a solar year. By 1582, those 11 minutes each year had added up to 10 days. Pope Gregrory XIII aligned the calendar with the solar year by declaring October 5, 1582, to be October 15, 1582, in the Catholic regions of the world, thus correcting for the 10 lost days.

What is the **Gregorian calendar**?

In addition to moving the calendar forward by 10 days in 1582, Pope Gregory XIII also corrected the error of the Julian calendar. He declared that

Why do we have leap years?

We have leap years to keep the calendar accurate with respect to the solar year. It takes the Earth 365 days, 5 hours, 48 minutes, and 46 seconds—just under 365.25 days—to revolve around the sun. If there were no leap years, every 56 years the calendar would be two weeks behind. By adding one extra day every four years, the calendar stays accurate.

years ending in "00" would not be leap years, except those divisible by 400 (such as the year 2000). The Julian calendar, with Pope Gregory's correction, is known as the Gregorian calendar, which most of us use today.

When was the **Gregorian calendar adopted in the United States**?

Though Catholic countries switched to the Gregorian calendar in the 16th century, Protestant countries, such as England and its colonies, refused to switch from the Julian to the Gregorian calendar at that time. It wasn't until 1752 that Britain and its colonies, including the colony that soon thereafter became the United States, switched to the Gregorian calendar. By the that time, there was an 11-day difference in time, so September 3, 1752, became September 14, 1752.

Is the Gregorian calendar **accurate**?

Almost! It is still 25 seconds longer than the solar year. Therefore, after about 3,320 years we will be a full day ahead of the solar year. The keepers of time will have to deal with this problem when the time comes.

What type of calendar did the **French** use between 1793 and 1806?

In 1793, during the French Revolution, an entirely new calendar was established by the National Convention. The calendar was designed to

153

help rid French society of its Christian influences. Within this new calendar there were 12 months, each consisting of three décades. Each décade was composed of 10 days. Five days (six during leap year) were added at the end of the year to add up to 365 (or 366) days. Napoleon reinstated the Gregorian calendar in 1806.

What type of calendar was used in the **Soviet Union** between 1929 and 1940?

The Soviets created the Revolutionary calendar, which had 5 days in a week (four for work, the fifth as a day off) and six weeks in a month. Five days (six during leap year) were added at the end of the year to add up to 365 (or 366) days.

Which **celestial bodies** are the days of the week named after?

The names of the days of the week come from Roman or Norse names for the planets:

Day	Celestial Body (Roman/Norse)
Sunday	Sun/Sol
Monday	Moon
Tuesday	Mars/Tui
Wednesday	Mercury/Woden
Thursday	Jupiter/Thor
Friday	Venus/Frygga
Saturday	Saturn

When does the **21st century** begin?

The 21st century will begin at 12:00 a.m. January 1, 2001. Since the first century, which spanned the years 1 to 100, centuries have been counted beginning with the year ending in "01" rather than "00." For instance, the 20th century consists of the years 1901 through 2000.

**Where on the planet is it
light 24 hours a day in the summer?**

In the extreme north and south parts of the Earth (north of 66.5 degrees north, and south of 66.5 degrees south latitudes) it is light 24 hours a day during the summer and dark 24 hours a day during the winter.

THE SEASONS

How does the **tilt of the Earth** affect the seasons?

Since the Earth is tilted 23.5 degrees, the sun's rays hit the northern and southern hemispheres unequally. When the sun's rays hit one hemisphere directly, the other hemisphere receives diffused rays. The hemisphere that receives the direct rays of the sun experiences summer; the hemisphere that receives the diffused rays experiences winter. Thus, when it is summer in North America, it is winter in most of South America, and vice versa.

What are the **Tropics of Cancer and Capricorn**?

The two Tropics are the lines of latitude where the sun is directly overhead on the summer solstices. The Tropic of Cancer is at 23.5 degrees north and passes through central Mexico, northern Africa, central India, and southern China. The Tropic of Capricorn is at 23.5 degrees south and passes through central Australia, southern Brazil, and southern Africa.

What are the **solstices**?

There are two solstices—one on June 21 and the other on December 21. On June 21, the sun is directly above the Tropic of Cancer at noon and

155

heralds the beginning of summer in the Northern Hemisphere and the beginning of winter in the Southern Hemisphere. On December 21, the sun is directly above the Tropic of Capricorn at noon and heralds the beginning of winter in the Northern Hemisphere and the beginning of summer in the Southern Hemisphere.

Can you stand an **egg on end** only on the spring equinox?

It is a common legend that an egg can be balanced on its end only on the spring equinox (March 21). Actually, there's nothing magical about gravity on the spring equinox that would allow an egg to stand on end—it can happen at any time of the year with patience and perseverance.

What are the **equinoxes**?

There are two equinoxes—one on March 21 and the other on September 21. On both equinoxes, the sun is directly over the equator. March 21 heralds the beginning of spring in the Northern Hemisphere and the beginning of fall in the Southern Hemisphere. September 21 heralds the beginning of fall in the Northern Hemisphere and the beginning of spring in the Southern Hemisphere.

Where are the **Arctic and Antarctic Circles**?

The Arctic Circle is located at 66.5 degrees north of the equator and the Antarctic Circle is located at 66.5 degrees south of the equator. Areas north of the Arctic Circle and south of the Antarctic Circle have 24 hours of light during the summer and 24 hours of darkness during the winter.

EXPLORATION

EUROPE AND ASIA

Who **disguised himself as a Muslim** to travel to Mecca?

Since non-Muslims are not allowed into the sacred city of Mecca, British explorer Sir Richard Francis Burton disguised himself as an Afghan pilgrim in order to enter the city in 1853. Burton, having learned various languages in the military, explored India, the Middle East, Africa, and South America. Also a prolific writer, Burton published many accounts of his journeys and is perhaps most famous for translating *1001 Arabian Nights* into English.

What were **Marco Polo's** contributions to exploration?

Though Marco Polo did not actually discover anything, his writings in *Travels of Marco Polo* served as Europe's introduction to the East, and spurred interest in exploration. Marco Polo, born in the mid-13th century in Venice, traveled with his father and uncle to China. During his stay, Polo served the Emperor Kublai Khan as an ambassador, as a governor, and in a host of other diplomatic positions. In his 30s he returned to Venice and fought against the city-state of Genoa and was eventually captured. While imprisoned in Genoa, he dictated the story of his travels to a fellow prisoner, creating the somewhat exaggerated memoir *Travels of Marco Polo*.

Which explorer was named the **Grand Imperial Eunuch** by the Emperor of China?

The Chinese explorer Cheng Ho helped Emperor Yung-lo come to power in 1402, and in 1404 the Emperor named Cheng Ho the Grand Imperial Eunuch. In 1405, Cheng Ho set sail on the first of his seven voyages, which spread Chinese influence and knowledge throughout South Asia and Africa. China moved toward isolationism after Yung-lo died.

Who was **Alexander von Humboldt**?

Alexander von Humboldt (1769–1862) was a German geographer who explored South America, Europe, and Russia. Von Humboldt traveled deep within the Amazon rain forests, developed the first weather map, and wrote a five-volume encyclopedia in which he sought to describe all of human knowledge about the Earth. Von Humboldt was the world's last great polymath (one of encyclopedic learning).

Who was the greatest **explorer of the Arab world**?

Known as the "Muslim Marco Polo," Ibn-Batuta (1304–1369) explored much of Africa and Asia. In his lifetime, Ibn-Batuta traveled over 75,000 miles, gaining the reputation of the most-traveled man on Earth.

What did **Genghis Khan** conquer?

Genghis Khan, the ruler of the Mongol Empire, conquered an area stretching from China to western Russia to the Middle East. Khan created the world's largest empire, which began to dissolve following his death in 1227.

Who was **Prester John**?

In the 12th century, a letter arrived for the Pope claiming to be from "Prester John," the leader of a Christian kingdom in the east that was in danger of being overrun by infidels. Prester John reportedly asked for help from European brethren. Though Prester John and his kingdom were never discovered, his mysterious letter sparked travels and explorations for centuries in an attempt to rescue the kingdom.

What were the **Crusades**?

From the 11th through the 14th centuries, groups of armed Christian Europeans invaded the Middle East to take the Holy Land from the Muslims and reclaim it for Christianity. The Crusaders ruthlessly murdered and pillaged throughout their long journey to the Middle East, and continued their brutality once there. Though the Crusades were a horrific era, the knowledge of the world gained by the Crusaders spurred a better geographic understanding.

AFRICA

Who discovered the **source of the Nile River**?

In 1856, John Hanning Speke was sent by the British Royal Geographical Society to discover lakes believed to exist in eastern Africa. In 1858, Speke and fellow explorer Sir Richard Francis Burton discovered Lake Tanganyika. Speke and Burton split, and while traveling on his own,

159

Which explorer fought for both the Union and Confederate armies in the U.S. Civil War?

Though born in Britain, Sir Henry Morton Stanley sailed to the United States and worked there for several years before the start of the Civil War. He joined the Confederate Army but was captured in 1861 at the Battle of Shiloh. He then joined the Union Army. Stanley is best known for his search for the missing explorer, David Livingstone, and his greeting upon finding him: "Dr. Livingstone, I presume?"

Speke discovered Lake Victoria and claimed it to be the source of the Nile River. Though many did not believe the lake was the source, Speke returned to the lake in 1860 and proved Lake Victoria was indeed the source of the Nile River.

Whose **body was preserved** and then hand-carried for nine months to the coast of Africa?

Dr. David Livingstone, the world-famous explorer of Africa, died while exploring the area now known as Zambia. His body was embalmed with sand and his heart was buried under a nearby tree. His body was then wrapped in cloth and covered with tar to waterproof it. Loyal servants carried his body for nine months, all the way to the eastern coast of Africa, where the body was then transported to Britain on the HMS *Vulture*. On April 18, 1874, his body was buried in Westminster Abbey.

Who was **Captain Kidd**?

Though Captain William Kidd was hired by the British to fight pirates, he himself soon became a pirate. After he set sail for Africa, an area swarming with pirates, reports returned to Britain that Kidd himself had

Leif Ericson, off the coast of Vinland (Newfoundland). (Corbis–Bettmann)

captured several ships. Upon learning of a warrant for his arrest, Kidd sailed to Boston to meet a benefactor, who then had Kidd arrested and sent back to Britain. On May 23, 1701, Kidd was hanged for piracy.

THE NEW WORLD

Who was the **first European** to reach **North America**?

In the early 11th century, Leif Ericson was the first European to set foot in North America. According to legend, Ericson, a Norse explorer, visited Helluland, Markland, and Vinland, which are believed to be Baffin Island, Labrador, and Newfoundland, respectively.

Who was **Ponce de Leon**?

The Spanish conquistador Juan Ponce de Leon searched for the fountain of youth. In 1513 during his travels in search of the mythical fountain,

Christopher Columbus in the New World. (Archive Photos)

reportedly located on the legendary island of Bimini, Ponce de Leon discovered Florida.

Who was the first European to see the **Pacific Ocean** from its eastern shore?

In 1513, Vasco Núñez de Balboa crossed the Isthmus of Panama and became the first European to see the Pacific Ocean from its eastern side. Wearing full armor, he walked straight into the ocean and claimed it and all the land it touched for Spain. He named the discovery the "Mar del Sur" (the South Sea). Only six years later, Balboa was beheaded by a jealous rival.

What is the **Strait of Magellan**?

Ferdinand Magellan discovered the strait that bears his name in 1520 during his voyage that circumnavigated the Earth. Magellan used the strait as a shortcut around the southern tip of South America. The

waters of the Strait of Magellan are violent and surrounded by dangerous rocks.

What was the intent of **Magellan's Expedition**?

Ferdinand Magellan left Europe in 1519 hoping to circumnavigate the
globe. He succeeded in reaching the Phillippines in 1521, where he was
later killed by natives. Though five ships and 241 men left Europe on
September 20, 1519, only one ship, *The Victoria*, returned to Spain with
18 men on September 6, 1522. Despite Ferdinand Magellan's death on
April 27, 1521, during a war in the Philippines, the Magellan Expedition
successfully circumnavigated the globe.

Why is **Columbus** credited with discovering the New World?

Despite the fact that Christopher Columbus was not the first person nor
the first European to reach the Americas, Columbus's discovery was
important in that it prompted mass exploration and colonization of the
New World. Though the idea of Columbus "discovering" the New World
is extremely Eurocentric, Europeans have credited Columbus with the
discovery of, and the enthusiasm surrounding the exploration of, the
New World.

Did everyone during Columbus' time think that the **world was flat**?

No, they did not. Though the common perception is that Christopher
Columbus had to convince King Ferdinand and Queen Isabella that
the world was a sphere, Columbus actually had to convince the King
and Queen of the circumference of the world. Though the ancient
Greeks had discovered that the Earth was a sphere, centuries passed
before this was generally accepted. By the 15th century, however—the
time of Columbus's sailing—most educated people believed the world
to be round. But the question remained: how far was it to travel
around the world?

Is it true that Columbus deliberately **fudged the circumference measurement** so as to make a better case for his trip?

Though most scholars believed that the circumference of the Earth was approximately 25,000 miles, Columbus used an estimate of 18,000 miles to push his case, in order to make his trip seem more achievable and the costs more reasonable. Columbus used Posidonus' smaller estimate, rather than Eratosthenes' larger and more accurate estimate, to make the trip appear shorter.

What did the **Mason-Dixon line** originally divide?

While the Mason-Dixon line commonly refers to the division between the "North" and "South" in the eastern United States, it was originally a boundary between Pennsylvania and Maryland surveyed by Charles Mason and Jeremiah Dixon in 1763. During the Civil War, the boundary between Pennsylvania and Maryland was extended westward to represent the line between the slave and non-slave states.

Who was **Vancouver, Canada,** named after?

In the 1790s, George Vancouver, who had previously accompanied James Cook on his explorations for "Terra Australis Incognito" and the Northwest Passage, explored and mapped the Pacific coast of North America. Vancouver circumnavigated Canada's Vancouver Island, and it and the city of Vancouver (founded 1881) were named for him.

What were **Lewis and Clark** looking for?

President Thomas Jefferson sent Merriwether Lewis and army officer William Clark to search for a Northwest Passage, a waterway that would connect the Atlantic and Pacific Oceans. Beginning in May 1804 and lasting through September 1806, the two men and their expedition party traveled through the uncharted Louisiana Territory and the Oregon Territory. Though they did not locate a Northwest Passage, Lewis and Clark documented the geography of the West.

Was George Washington a geographer?

George Washington manifested the abilities of a cartographer and a surveyor at an early age. When he was 13, Washington made his first map, which was of his father's property, Mount Vernon. At the age of 17, Washington was appointed surveyor of Culpeper County, Virginia. At age 21, Washington entered military service, and the rest is history.

Who was **John Wesley Powell**?

Though he lost an arm in the Civil War, John Wesley Powell became one of the leading surveyors of the 19th century. In 1869, Powell explored the Grand Canyon. Traveling on a boat along the Colorado River, he faced dangerous rapids, hostile Native Americans, and weather extremes. In 1880, Powell was appointed the second director of the United States Geological Survey.

How did **America** get its **name**?

Though Christopher Columbus discovered the New World, he always believed that he had reached Asia, not realizing that he had encountered new continents. The Italian explorer Amerigo Vespucci, who had explored the New World and published accounts of his travels, was the first person known to have distinguished the New World from Asia. The German cartographer Martin Waldseemüller, who had read of Amerigo Vespucci's travels, published a map of the New World in 1507 with what is now known as South America named "America," in honor of Amerigo. The name stuck.

What did **James Cook** not discover?

During the 18th century, James Cook was sent on several expeditions of discovery, one of which was to the southern Pacific Ocean in search of the legendary landmass "Terra Australis Incognita." Though the conti-

James Cook, English explorer and navigator. (Archive Photos)

nent that is now known as Australia had already been discovered, a centuries-old belief foretold of another huge continent in that area. Cook traveled to the southern Pacific Ocean and disproved the legend of "Terra Australis Incognita." On another expedition, Cook was sent to find a water route north of North America from Asia to Europe. As Cook sailed, he discovered the Sandwich Islands (Hawaiian Islands) and determined that a Northwest Passage was not feasible because of ice. On his way back from this "un"-discovery, Cook was killed in the Sandwich Islands during a struggle over the theft of one of his boats.

How fast did the **Mayflower** sail?

In 1620, the Pilgrims sailed on the Mayflower from Plymouth, England, to the New World in 66 days. Though the Mayflower relied upon intermittent wind for propulsion, it averaged two miles per hour across the Atlantic Ocean.

What is a **nautical mile**?

Used for measuring ocean-based distances, a nautical mile is equivalent to approximately 6,076 feet or 1.15 miles. The speed of ships is measured in knots. One knot is equivalent to one nautical mile per hour.

What **monetary unit** is used in Panama?

The currency in Panama is called the Balboa, after the explorer Vasco Núñez de Balboa, because Balboa established the first European settlement in Panama.

THE POLES

Who was the first person to reach the **North Pole**?

Though American explorer Robert Edwin Peary is credited as the first to reach the North Pole, it is likely that he only came within 30 to 60 miles

Robert Scott's party at the South Pole. (Hulton-Deutsch Collection/Corbis)

of 90° North during his expedition in 1909. Who did actually reach the Pole first is still being debated.

Who was the first person to reach the **South Pole**?

In 1911, the Norwegian explorer Roald Amundsen and the British explorer Robert Scott were racing against each other to be the first to reach the South Pole. On December 4, 1911, Amundsen and his crew of four reached the South Pole at 90° South. Approximately one month later, Scott and his team arrived at the pole. Depressed from their defeat and with inadequate supplies of food, Scott and his team died while trying to return to their base camp.

UNITED STATES OF AMERICA

PHYSICAL FEATURES AND RESOURCES

Where is the **center** of the contiguous United States?

The geographic center of the lower 48 states is located at 39° 50' north, 98° 35' west, approximately four miles northwest of Lebanon, Kansas.

What is the **highest point** in the United States?

Alaska's Mt. McKinley (also known as Denali) is the highest point in the United States at 20,320 feet. In the contiguous 48 states, the highest point is California's Mt. Whitney at 14,495 feet, which is less than 100 miles from North America's lowest point, Death Valley (282 feet below sea level).

What is the highest point **east of the Mississippi**?

North Carolina's Mt. Mitchell is the tallest point east of the Mississippi River at 6,684 feet.

What is the **highest lake** in North America?

Yellowstone Lake in Yellowstone National Park is the highest lake in North America, at 7,735 feet above sea level.

169

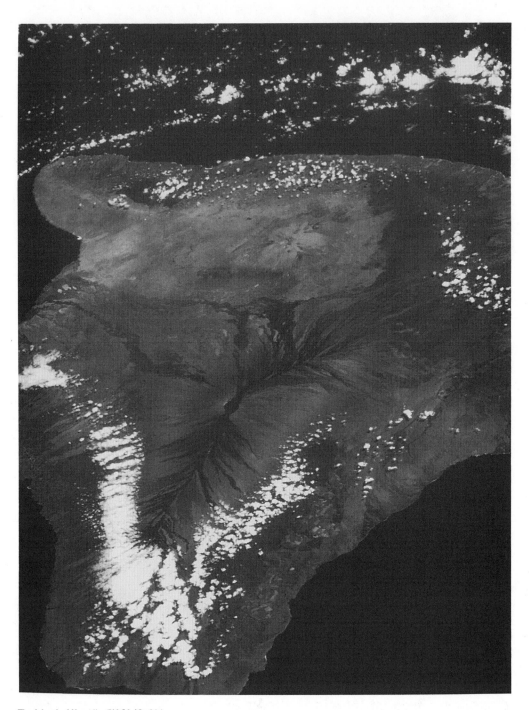

The Island of Hawaii. (NASA/Corbis)

What is the **deepest lake** in the United States?

Crater Lake in Oregon, lying within the collapsed crater of an ancient volcano, is the nation's deepest lake at 1,932 feet. Crater Lake has no feeder streams—it is solely filled by precipitation.

What is the **largest island in the United States**?

The Island of Hawaii is the largest island in the United States at 4,021 square miles. Puerto Rico is the second largest at 3,435 square miles.

Where is the world's **largest marsh**?

The Everglades in Florida is the world's largest marsh, consisting of 2,185 square miles. The water across this southern Florida marsh averages six inches in depth. The Everglades is an endangered ecosystem, threatened by excess drainage and the introduction of exotic plants.

Where is the world's **largest mountain**?

Hawaii's Mauna Kea is the world's largest mountain. It begins on the sea floor and rises 33,480 feet (Mount Everest only rises 29,000 feet). Mauna Kea's peak reaches 13,796 feet above sea level.

How many **Great Lakes** are there?

There are five Great Lakes: Huron, Ontario, Michigan, Erie, and Superior. The acronym "HOMES" can help you to remember the names of the five Great Lakes. All of the lakes except Michigan lie on the U.S.–Canada border.

What is the **largest freshwater lake** in the world?

Lake Superior is the largest freshwater lake in the world. It is 31,700 square miles in area and is approximately 350 miles long.

171

Where do the Great Lakes rank in terms of size?

Lake Superior is 31,700 square miles and is the world's largest freshwater lake; Lake Huron is 23,000 square miles and the third largest freshwater lake; Lake Michigan is 22,300 square miles and the fourth largest freshwater lake; Lake Erie is 9,900 square miles, the world's tenth largest freshwater lake; and Lake Ontario, 7,300 square miles, is the twelfth largest freshwater lake.

What is the world's **shortest river**?

The world's shortest river is Oregon's D River, which is a mere 120 feet long. It connects Devil's Lake to the Pacific Ocean near Lincoln City, Oregon.

Why is **Coney Island** called an island even though it's not?

Though now a peninsula of Long Island, Coney Island was actually an island at one time. The popular amusement park of the early 20th century is now attached to Long Island due to the silting up of Coney Island Creek, which once separated the two islands.

THE STATES

What are the five **largest states**?

The five largest states are Alaska (591,000 square miles), Texas (266,800 square miles), California (158,700 square miles), Montana (147,000 square miles), and New Mexico (121,600 square miles).

What are the five **smallest states**?

The five smallest states are Rhode Island (1,200 square miles), Delaware (2,000 square miles), Connecticut (5,000 square miles), Hawaii (6,500 square miles), and New Jersey (7,800 square miles).

What are the five **most populous states**?

As of the 1990 Census, the five most populous states were California (30 million), New York (18 million), Texas (17 million), Florida (13 million), and Pennsylvania (11.9 million). By the middle of the 1990s, estimates showed that Texas had a greater population than New York—proof will come in the 2000 Census.

What are the five **least populous states**?

As of the 1990 Census, the five least populous states were Wyoming (450,000), Alaska (550,000), Vermont (563,000), North Dakota (639,000), and Delaware (666,000).

Which state has the **most lakes**?

Though Minnesota is known for its "10,000 lakes" (the slogan on its license plates), it is not the state with the most lakes. Neighboring Wisconsin has even more, about 14,000 lakes, but Alaska is the definite winner, with over three million lakes!

How many states have **four-letter names**?

There are three states that tie for having the shortest name: Utah, Ohio, and Iowa.

How many states **end in the letter "a"**?

Twenty-one of the 50 states end in the letter "a."

Why is Rhode Island called an island even though it's not?

Rhode Island's official name is "Rhode Island and Providence Plantations," and includes not only the area of the mainland (the location of the city of Providence), but also four major islands. The largest of these islands is named Rhode Island, from which the state gets the first part of its name.

How was the **Delmarva Peninsula** named?

The Delmarva Peninsula, located on the East Coast, contains all of Delaware and portions of Maryland and Virginia. The 180-mile long peninsula was named by combining the abbreviations of each of those three states—Del., Mar., and Va.

How many **islands** does **Hawaii** include?

There are a total of 122 islands in the Hawaiian chain. The southernmost island, Hawaii, is the largest (4,000 square miles). The two westernmost islets are known as the Midway Islands and, while they are part of the United States, they are not part of the state of Hawaii.

Which Hawaiian island was once a **leper colony**?

The island of Molokai, located between Maui and Oahu, served as an isolated leper colony from 1868 until the 1940s. Leprosy, a degenerative disease that attacks the skin and nerves, became epidemic in Hawaii in the mid-19th century. The living conditions on Molokai were deplorable until the arrival of missionaries in 1873. In the 1940s, new drugs made leprosy treatable, rendering the leper colony on Molokai no longer necessary. There is now a leprosy treatment center on the island.

Which state has the highest divorce rate?

Nevada's divorce rate is double nearly all other states, with a rate of 14 divorces per thousand people. For over a century, Nevada had been renowned as an easy place to obtain a divorce. Even now, many people travel to Nevada and fill Nevada's brief residency requirement of six weeks to obtain a divorce.

What do the **Sandwich Islands** and Hawaiian Islands have in common?

The Sandwich Islands and Hawaiian Islands are actually the same set of islands. In 1778, when Captain James Cook discovered the islands, he named them the Sandwich Islands. Gradually, the islands began to be known by their indigenous name of Hawaii. Cook named the islands after his supporter, John Montagu, the fourth Earl of Sandwich.

What is the only state with a **diamond mine**?

Arkansas is the only state with a diamond mine. Located in southwestern Arkansas, the mine is no longer in commercial production but is now a park, Crater of Diamonds State Park, that allows visitors to take any diamonds they find.

What is the **driest state**?

Nevada averages only 7.5 inches of rainfall annually, making it the driest state in the union.

What is the **official neckwear of Arizona**?

Like an official flower or animal, each state can add additional official symbols for their state. The official neckwear of Arizona is the bolo tie.

Which state **borders only one other state**?

Bordering New Hampshire, Maine is the only state that borders just one other state.

Which state **borders the most** others?

Tennessee borders eight states: Arkansas, Missouri, Kentucky, Virginia, North Carolina, Georgia, Alabama, and Mississippi.

Which state has the **longest coastline**?

Alaska's coastline, over 5,500 miles long, is longer than any other state's. California is second with 840 miles of coast.

What state has only **one legislative body**?

Nebraska has a unicameral legislature; all other 49 states have two houses.

How many states have land **north of Canada's** southernmost point?

Twenty-seven of the 50 states have land north of Canada's southernmost point, Pelee Island in Lake Erie.

CITIES AND COUNTIES

How many cities are in the United States?

There are approximately 19,300 incorporated cities in the United States. A vast majority—about 18,000—have less than 25,000 people.

St. Augustine, Florida, 1753. (Library of Congress/Corbis)

Measured by area, what is the **largest city** in the United States?

Juneau, Alaska, the capital of the state, is the largest city in the United States, with over 3,100 square miles.

What is the **fastest-growing metropolitan area** in the United States?

From 1990 to 1994, the Las Vegas metropolitan area population increased 26.2 percent (from 853,000 to 1,076,000 people), making it the fastest-growing metropolitan area in the United States. Las Vegas, located in Nevada near the California border, has been an entertainment and gambling center for decades.

What are the **northernmost, southernmost, easternmost, and westernmost** cities in the United States?

Barrow, Alaska, is the northernmost; Hilo, Hawaii, is the southernmost; Eastport, Maine, is the easternmost; and Atka, Alaska, is the westernmost city in the country.

What is the **highest settlement** in the United States?

Climax, Colorado, is located at an elevation of 11,302 feet and is the highest settlement in the United States.

What is the oldest continually **occupied city** in the United States?

Saint Augustine, Florida, was founded in 1565 by Spanish explorer Pedro Menendez de Aviles. Saint Augustine is located on Florida's eastern (Atlantic) coast and now has a population of approximately 12,000. It is not only the oldest continually occupied city in the United States, but also in all of North America.

Which is **further west**—Los Angeles, California, or Reno, Nevada?

Though Nevada is California's eastern neighbor and Los Angeles sits on the Pacific Coast, Reno is farther west than Los Angeles. Reno is located at 119°, 49' west, while Los Angeles is located at 118 degrees, 14' west.

What city is **named for a game show**?

The popular radio game show "Truth or Consequences" offered to host its 10th anniversary show in a city that would change its name to the show's title. In 1950 Hot Springs, New Mexico, changed its name and the 10th anniversary show was broadcast from Truth or Consequences, New Mexico. The city still holds the name today.

What city is known as the **Earthquake City**?

Charleston, South Carolina, claims the nickname "Earthquake City." On August 31, 1886, Charleston suffered from the largest earthquake in history to strike the east coast of the United States. Sixty were killed in the quake, which had an estimated Richter magnitude of 6.6.

Where does **Los Angeles** get its water?

Not much of the water in Los Angeles comes from local sources—most of it is brought from hundreds of miles away. Large aqueducts, man-made channels used to transport water, were built to carry water from Owens Valley (in East-Central California), from the Colorado River, and from the rivers of Northern California, to Los Angeles. Though this method has brought fresh water to a region that desperately needs it, it has also drained and damaged the ecologies that once depended on the water now being sapped from its supply.

How many **counties** are in the United States?

There are 3,043 counties in the United States. The state with the least number of counties is Delaware, with three, and Texas has the most, with 254.

What are the **largest and smallest counties** in the United States?

San Bernardino County in California is the largest, at 20,000 square miles. It stretches from metropolitan Los Angeles to the Nevada/Arizona border. The smallest county in the United States is Kalawao, Hawaii, at 13 square miles, which has a very small population of only 130.

PEOPLE AND CULTURE

What is the **population** of the United States?

The United States' population has grown from 3.9 million in 1790 to 249 million in the 1990 Census. In 1998, the population estimate was 270 million. Projections indicate a population of 288 million by 2005, 313 million in 2015, and 338 million in 2025. In 1990, the annual percentage increase was one percent.

Where do most **legal immigrants** to the United States come from?

In 1993, over 126,000 legal immigrants came from Mexico; over 65,000 came from China; more than 63,000 came from the Philippines; and over 59,000 came to the United States from Vietnam.

What is the **oldest college** in the United States?

Harvard University is the oldest college or university in the United States. It was founded in 1636 in Cambridge, Massachusetts, just outside of Boston.

What is the most popular **national park** in the United States?

North Carolina and Virginia's Blue Ridge Parkway is the most popular national park, drawing over 17 million visitors each year. Located in the southern Appalachian Mountains, the Parkway is a highway through beautiful scenery and includes nearby hiking trails, ranger talks, and bird-watching. The other two most-visited national parks in the United States are Golden Gate National Recreation Area, which gets about 14 million visitors annually, and Lake Mead National Recreation Area, with 9.4 million.

What is the **oldest public park** in the United States?

In 1634, William Blackstone sold 50 acres of his land to the town of Boston. This pastureland was set aside for common use and called the

**How many visitors
come to the United States every year?**

Over 46 million people visit the United States from foreign countries each year. About half of the tourists come from Canada (15 million) and Mexico (9 million) combined. Japan is the third most plentiful source of tourists (5 million).

Boston Common, making it the oldest public park in the United States. The Common is situated in front of Massachusetts' State House and has always been open public land.

What is the most popular **theme park** in the United States?

Disney operates the country's most popular theme parks. Opened in 1971, the Magic Kingdom in Orlando, Florida, is the most popular theme park in the country and draws over 12 million visitors annually. The three next-most popular theme parks are Disneyland, EPCOT Center, and Disney-MGM Studios.

Where was the world's first **monument to an insect** established?

On December 11, 1919, Enterprise, Alabama, dedicated a monument to the boll weevil. This tall statue of a woman with raised arms holding a boll weevil declares, "In profound appreciation of the boll weevil and what it has done as the herald of prosperity." The boll weevil, a beetle that attacks bolls of cotton, spread across the South at the beginning of the twentieth century wiping out cotton crops. Residents of Enterprise switched from cotton crops to peanut crops, thus discovering a new era of prosperity. The monument to the boll weevil is to remind residents and visitors alike of the resourcefulness of the community and the ability of man to diversify.

Mount Rushmore, located in the Black Hills of South Dakota. (Archive Photos)

How many **automobiles** are there in the United States?

There are approximately 151 million automobiles in use in the United States, about one car for every two people.

Where was the **first commercial air flight**?

On January 1, 1914, the first scheduled commercial air flight took passengers from Tampa Bay, Florida, to St. Petersburg, Florida. The service, which lasted only a few weeks, took one person at a time over the 22-mile route.

Where is the **sunbelt**?

The sunbelt, known for its warm temperatures, is a geographical area that spreads across the southern and southwestern states of the United States. It has been an area of high population growth over the past few decades, as more families and individuals have moved to states like California, Arizona, Texas, and Florida.

Where is the **rustbelt**?

The rustbelt is a term used to describe the United States' declining manufacturing region of the northeast and Midwest. Factory closures, especially those of steel and textile mills, have resulted in massive unemployment and declining population in rustbelt cities. The rustbelt runs from about Minnesota to Massachusetts.

Where is the **Bible Belt**?

The Bible Belt, a region noted for its high proportion of fundamentalist Christian beliefs, is located in the southern and Midwestern United States, running from about Oklahoma to the Carolinas.

Who carved **Mount Rushmore**?

Gutzon Borglum, an American sculptor, designed this national memorial located in the Black Hills of South Dakota. Construction began in 1927 and was nearly complete when Borglum died in March, 1941. Borglum oversaw construction of the 60-foot-tall heads of Presidents George Washington, Thomas Jefferson, Abraham Lincoln, and Theodore Roosevelt. After Borglum's death, his son completed the work on the unfinished Roosevelt by the end of 1941.

Why was there a **Russian outpost** in California?

Established in 1812, Fort Ross was a Russian outpost located in what is now Sonoma County, California. The outpost was started so that Russian fur traders could explore and exploit the area. A Russian presence was maintained at the fort until 1841.

What is the **oldest continuously published newspaper** in the United States?

In 1764, Thomas Green founded the *Hartford Courant* in Hartford, Connecticut, which is the oldest continuously published newspaper in the United States.

Where was the **first shopping mall** in the United States?

In 1922, Country Club District opened in the suburbs of Kansas City, Kansas, and was the first shopping mall in the country.

What is the **leading cause of death** in the United States?

Approximately one-third of all Americans die from cardiovascular disease, making it the country's leading cause of death. Cancers are the second leading cause of death in America, killing over one-quarter of the population.

Where is **Acadiana**?

In 1755, the British took control of Acadie, New Brunswick, Canada, and exiled the city's inhabitants. Forced to leave their homes, these people, known as Acadians, moved to Louisiana. The Acadians are still present in Louisiana and their culture has produced famed food and music, called Cajun. The area in southern Louisiana where the Acadians still live is termed Acadiana.

Why are **graves above the ground** in Louisiana?

In Louisiana, the water level is so close to the surface of the ground that coffins, rather than being buried, are placed in tombs above ground to avoid the possibility of coffins floating out of place. A Louisiana cemetery looks like a miniature city, with tombs and alleyways.

HISTORY: CREATING THE UNITED STATES OF AMERICA

How did the **United States** reach its present form?

The United States began as 13 British colonies on the Atlantic Coast. In 1783, the U.S. gained the Northwest Territory, the area encompassing what is now Ohio, Indiana, Illinois, Michigan, and Wisconsin. Spain and the U.S. agreed on the northern boundary of Florida in 1798 and the U.S. then took control of the Mississippi Territory. In 1803, the Louisiana Purchase (which includes most of the area west of the Mississippi) doubled the size of the country. In 1845 the independent Republic of Texas was annexed and Spain ceded Florida to the U.S. In 1846, the Oregon Territory (which included Oregon, Washington, and Idaho) was officially designated with a treaty between the U.S. and the United Kingdom. The Mexican War of 1846–1848 led to the secession of California, Utah Territory, and New Mexico territory to the U.S. The Gadsden Purchase of 1853 added southern Arizona. Alaska was purchased from Russia in 1867 and, finally, Hawaii was annexed by the U.S. in 1898.

What was **Manifest Destiny**?

First used by John Louis O'Sullivan in an 1845 editorial, Manifest Destiny was the phrase used to describe the assumption that American expansion to the Pacific Ocean was inevitable and ordained by God. The phrase was used to defend the annexations of Texas, California, Alaska, and even Pacific and Caribbean islands.

What were **Lewis and Clark** looking for?

President Thomas Jefferson sent Merriwether Lewis and army officer William Clark to search for a Northwest Passage, a waterway that would connect the Atlantic and Pacific Oceans. Beginning in May 1804 and lasting through September 1806, the two men and their expedition party traveled through the uncharted Louisiana Territory and the Oregon Territory. Though they did not locate a Northwest Passage, Lewis and Clark documented the geography of the West.

What was **Seward's Folly**?

Seward's Folly, also known as Seward's Icebox, was the derogatory nickname given to the area known as Alaska, purchased by the United States from Russia in 1867. The $7.2 million purchase, heavily encouraged by Secretary of State William Seward, was criticized by many, thus dubbed

Merriwether Lewis and William Clark meeting with Native Americans on their search for a Northwest Passage. (Corbis-Bettmann)

"Seward's Folly." The Alaskan Gold Rush of 1900 proved Seward to be a very wise man. In 1959, Alaska became the 49th state.

When was the most **territory added** to the United States at one time?

In 1803, the United States purchased over 800,000 square miles of land from France for $15 million. This territory, known as the Louisiana Purchase, extended from the Mississippi River to the Rocky Mountains and doubled the size of the United States.

Aside from the Louisiana Purchase, how was the **American West** obtained?

There were several other purchases and wars fought to gain the land west of the Louisiana Purchase. These include the Gadsen Purchase from Mexico, sections of the west ceded to the U.S. by Mexico, Texas, and the Oregon Country.

What was the **Oregon Trail**?

The Oregon Trail was a 2,000-mile-long pioneer trail that extended from Independence, Missouri, to Portland, Oregon. Migrants traveled along this trail in an effort to reach and settle the sparsely populated American West. It took migrants approximately six months to traverse the Oregon Trail and reach Oregon. The trail was heavily used from the 1840s and subsequent decades. Portions of the Trail are still visible today in places such as the Whitman Mission National Historic Site in Washington.

What was the **Trail of Tears**?

In 1838, the United States rounded up approximately 15,000 members of the Cherokee Nation and forced them from their homes in Tennessee to a reservation in Oklahoma. The removal of the Cherokee Nation from their land was done so that citizens of the United States could use the fertile lands in Tennessee. Along the "trail," approximately one-fourth of

187

the Cherokees died from malnutrition, disease, and government inefficiency.

How did the **Monroe Doctrine** protect the Americas?

In 1823, President James Monroe gave a speech that declared the Americas off limits to European powers. These policies became known in the 1840s as the Monroe Doctrine. Since 1823, the United States has used the Monroe Doctrine not only to prevent intervention by Europeans but also to further its own expansionist goals.

How did the United States obtain the **Virgin Islands**?

The United States purchased the Virgin Islands, which includes three main islands (St. Croix, St. John, and St. Thomas) along with 50 smaller islands, from Denmark in 1917 to help defend the Caribbean Sea. Just under 100,000 people live on the islands, which are now a territory of the United States.

Is **Puerto Rico** a state?

No, Puerto Rico is a commonwealth of the United States. In 1898, the United States acquired the island of Puerto Rico from Spain, but Puerto

Rico did not become a commonwealth until 1952. Though Puerto Ricans pay no federal income tax and cannot vote for President, they are citizens of the United States and can move freely around the country. Puerto Rico is allowed one representative to the House of Representatives, but that member cannot vote. With approximately 3.6 million residents, Puerto Rico is currently the largest colony in the world. In 1999, Puerto Ricans may have the opportunity to decide whether they would like to become a state of the U.S., become an independent country, or remain a colony.

NORTH AMERICA

See also Chapter 12, United States of America

What is the **center of North America**?

The geographic center of North America (including Canada, the U.S., and Mexico) is located six miles west of Balta, North Dakota, at 48° 10' north, 100°10' west.

What's **NAFTA**?

In 1994, Canada, the United States, and Mexico entered into the North American Free Trade Agreement (NAFTA), which reduces tariffs and economic controls between the three countries.

What is the **continental divide**?

The continental divide is simply the line that divides the flow of water in North America. Precipitation east of the divide flows towards the Atlantic Ocean while precipitation west of the divide flows towards the Pacific Ocean. The divide follows the line of the highest ranges of the Rocky Mountains and is not a distinct mountain range.

The continental divide in the Rocky Mountains of Canada. (Lowell Georgia/Corbis)

GREENLAND AND
THE NORTH POLE REGION

Is **Greenland** really green?

In 982 C.E., Greenland was named by its first colonizer, Eric the Red. Having been banished from Iceland, Eric the Red established a colony on Greenland and gave it a pleasant-sounding name, in order to attract other colonists. In reality, Greenland is not very green. Though small coastal areas are habitable, most of the island is mountainous and covered by ice sheets. Greenland is an autonomous territory of Denmark.

What is the **northernmost landmass** in the world?

Cape Morris Jesup, located in Greenland's Peary Land, is the northernmost point of land in the world. The cape is at 83° 38' north.

Is there land at the **North Pole**?

Since the North Pole lies in the Arctic Ocean, which is mostly covered by a large icecap year-round, there is no land near the North Pole. Though there is no land, animals such as the polar bear live upon the icecap.

CANADA

How many provinces are in Canada?

Canada is divided into 10 quasi-autonomous provinces and two territories. The 10 provinces are Alberta, British Columbia, Manitoba, New Brunswick, Newfoundland, Nova Scotia, Ontario, Price Edward Island, Québec, and Saskatchewan. Canada's two territories are Yukon Territory and the Northwest Territories. In 1999, a third area located in northeastern Canada will become a territory called Nunavut.

Where are the **Prairie Provinces**?

The grassy, central Canadian states of Manitoba, Saskatchewan, and Alberta are called the Prairie Provinces.

What is **Nunavut**?

Nunavut, Canada's third territory as of April 1, 1999, will become home to Canada's indigenous people, the Inuit. This new territory will cover approximately one-fifth of Canada's land area, but will contain less than one percent of Canada's population, with just over 20,000 people.

What are the largest **Canadian urban areas**?

Toronto (4.4 million people), Montréal (3.4 million), and Vancouver (1.9 million) are Canada's largest urban areas.

When did Canada have a **transcontinental railway**?

After nearly a decade of setbacks, Canada completed its first transcontinental railroad in 1885. The Canadian Pacific Railway opened western Canada to settlement and greatly helped the city of Vancouver to grow.

How is northern Canada **gaining elevation**?

During the ice ages, thick ice sheets covered northern Canada, including Hudson Bay, pushing the continent down by the sheer force of the extreme weight of the ice. Ever since the melting of the ice sheets at the end of the ice age, the ground has been "springing" back by rising a few inches each year.

Why do they speak **French in Québec**?

Most of Québec speaks French because the French founded the city of Québec, one of the oldest cities in North America, in the 17th century. Québec served as the capital of the surrounding New France until the British took control of the territory in 1763. Even though it was ruled

Where is Canada's population concentrated?

Over half of Canada's population lies in southern Ontario and southern Québec in eastern Canada. This area, dubbed "Main Street," stretches from Windsor, Ontario, to Québec City, Québec, and includes the major cities of Toronto, Ottawa, and Montréal.

by the British from the 18th century onward, the region that was once held by France has remained a center of French culture and language in North America.

The differences in culture between the province of Québec and the rest of Canada have been so extreme that there are strong secessionist forces within Québec that want Québec to become its own country. Though two referendums for Québec's independence have been voted upon in Canada-wide elections, both the 1980 and 1995 referendums failed to pass. But these failures have not quieted Québec's secessionist forces, and it is likely that future referendums will be held. If passed, many believe that the secession of Québec would dissolve the loose Canadian federation.

What did the phrase **"fifty-four forty or fight"** mean?

In the mid-19th century, many Americans wanted to see the United States' territory expand northward into the area now known as Canada. The phrase "fifty-four forty or fight" referred to the desire to move the boundary northward to 54° 40' north, which would have encompassed much of southern Canada. Ultimately, the boundary was fixed at 49° north, where it sits today.

What is the **deepest lake** in North America?

Northwestern Canada's Great Slave Lake is the continent's deepest, at 2,015 feet. The lake is named after the indigenous people who live near the lake, the Slave.

How long are the **Rockies**?

The Rocky Mountains (Rockies) extend over 2,000 miles, from the Yukon Territory in Canada to Arizona and New Mexico in the southern United States.

How was **Niagara Falls stopped**?

The water from the Niagara River falls over two waterfalls, divided by Goat Island. Only six percent of the water from the Niagara River falls over American Falls, while Horseshoe Falls (or Canadian Falls) carries the majority of the water. In 1969, a temporary dam was built to divert the water from American Falls to Horseshoe Falls for several months in order to study the erosion endemic to both waterfalls.

What was the **Klondike Gold Rush**?

In 1896, gold was discovered in an area of western Canada known as the Klondike, located in the Yukon Territory where the Klondike and Yukon Rivers meet. Once the news of the discovery spread, tens of thousands of people headed west, creating the Klondike Gold Rush.

What is the world's largest gulf?

The Gulf of Mexico, which is approximately 600,000 square miles, is the world's largest gulf. The Gulf of Mexico is surrounded by Mexico, the southern coast of the United States, and Cuba, and is connected to the Caribbean Sea in the east.

MEXICO

What was the **first major city** in the Western Hemisphere?

From the first through the seventh centuries, the city of Teotihuacan flourished. Located northeast of modern Mexico City, Teotihuacan had a maximum population of approximately 200,000 and was the first major city in the Western Hemisphere. The city was graced with the Pyramid of the Sun and the Pyramid of the Moon. Teotihuacan should not be confused with the Aztec city of Tenochtitlan, which was built nearly six centuries later in the area that is now known as Mexico City.

Where are **Sierra Madre Occidental and Sierra Madre Oriental**?

Sierra Madre Occidental and Sierra Madre Oriental are two mountain ranges in Mexico. The ranges' names stem from the meanings of "occidental" (western) and "oriental" (eastern); thus the Sierra Madre Occidental lies along Mexico's west coast and the Sierra Madre Oriental lies along the east coast.

How many states are in Mexico?

Mexico is divided into 31 states plus the federal district of Mexico City. The largest state is Chihuahua, located in northern Mexico, with 95,400 square miles. The most populous state is the state of Mexico, located west of Mexico City, with 10 million people.

Where is the **Yucatan**?

The Yucatan is a large peninsula in southern Mexico. The Yucatan Peninsula "points" toward Florida and separates the Gulf of Mexico from the Caribbean Sea.

How much of **Mexico's population** lives in Mexico City?

Approximately one-fourth of Mexico's population lives in Mexico City's metropolitan area. Mexico City, Mexico's capital, is the largest metropolitan area in the Western Hemisphere, with a population of 24 million.

How do **maquiladoras** help clothe the United States?

Maquiladoras are Mexican factories owned by foreign (usually U.S.) corporations. Most often located along the Mexican–U.S. border, maquiladoras receive raw materials from the United States and produce finished goods for export. Maquiladoras commonly produce clothing and automobiles.

CENTRAL AMERICA

What is the difference between **Central and Latin America**?

Central America includes the countries that connect North and South America and are located between Mexico and Colombia. The seven countries of Central America are Guatemala, Belize, El Salvador, Honduras, Nicaragua, Costa Rica, and Panama. Latin America is a much broader term, and includes Central America as well as Mexico and all of the countries of South America.

Construction of the Panama Canal in 1890. (Archive Photos)

What was the last Central American country to obtain its **independence**?

In 1981, Belize, formerly known as British Honduras, became the last North American country to obtain independence.

Are there lots of mosquitoes on the **Mosquito Coast**?

The Mosquito Coast is an area 64 miles wide and approximately 250 miles long along Nicaragua's eastern shore. Though the Mosquito Coast receives an average of 250 inches of rain annually, making it a perfect breeding place for mosquitoes, the Mosquito Coast was named after the indigenous people of the area, the Mosquito Indians.

Who owns the **Panama Canal**?

In 1903, the U.S.–backed revolutionaries in western Colombia revolted and created the independent country of Panama, which was immediately

199

recognized by the United States. The newly independent Panama gave the United States use of a 10-mile-wide strip of land across the Isthmus of Panama, where the U.S. built the Panama Canal. The United States maintained control of the Panama Canal and its surrounding land, called the Canal Zone, until 1979, when unrest in Panama over U.S. presence led to the return of the Canal Zone to Panama. The Panama Canal is also destined to return to the control of Panama in the year 2000.

Who is a **peon**?

A peon is a farm laborer in Central America who works on large farms known as haciendas.

What is the world's **second-longest barrier reef**?

The second-longest barrier reef in the world lies just off the Atlantic coast of Belize, on the northeastern corner of Central America, and consists of Lighthouse Reef and Glovers Reef. Belize's reefs are only a few dozen miles long while the Great Barrier, the longest reef in the world, is hundreds.

THE WEST INDIES

Where are the **East and West Indies**?

The East and West Indies are separated by half the planet. The West Indies are islands in the Caribbean, including the Greater Antilles, the Lesser Antilles, and the Bahamas; while the East Indies include Southeast Asia, India, and Indonesia. When Christopher Columbus reached the New World in 1492, he believed that he had actually found a shorter route to the East Indies. Thus, Columbus thought the islands he had reached made up a portion of the Indies and considered the islands' inhabitants to be "Indians."

> ## What is the difference between the Greater and Lesser Antilles?
>
> The Greater Antilles refers to the four largest Caribbean islands: Cuba, Hispaniola, Puerto Rico, and Jamaica. All smaller Caribbean islands make up the Lesser Antilles.

Where are the **Windward Islands**?

The Windward Islands are located in the Caribbean and are exposed to the northeast trade winds (northeasterlies) of the Atlantic Ocean. Because of their vulnerability to these winds, the islands were named the Windward Islands. The Windward Islands include Martinique, St. Lucia, St. Vincent, the Grenadines, and Grenada.

Where are the **Leeward Islands**?

The Leeward Islands are also located in the Caribbean and are less exposed to the northeasterlies. Because these islands are "lee," or away from the wind, they were named the Leeward Islands. The Leeward Islands include Dominica, Guadeloupe, Montserrat, Antigua, Barbuda, St. Kitts, Nevis, Anguilla, and the Virgin Islands.

Was **Cuba** ever a part of the United States?

The United States went to war against Spain in 1898 to assist Cubans who were rebelling against Spanish rule. The U.S. took control of Cuba during the Spanish-American War in 1898 and held it until 1902, when Cuba was granted independence. The three-year military occupation by the U.S. ended with an agreement that the United States would be allowed to lease Guantanamo Bay, which the U.S. still uses as a naval base.

How did Cuba become a **Communist** country?

Having been an independent country for 57 years, the Cuban government, run by the dictator Fulgencio Batista y Zalvidar, fell to the Communist leader Fidel Castro in 1959. Because of Cuba's Communist government, the United States severed its relationship with Cuba, forcing the island to ally itself with the Soviet Union. In October, 1962, the presence of this nearby Communist country caused extreme terror in the U.S. when the U.S.S.R. attempted to place nuclear missiles within Cuba. The "Cuban Missile Crisis" is thought to be the closest the Cold War ever got to a real nuclear war.

Where is the **Bay of Pigs**?

The Bay of Pigs is a bay in southwestern Cuba. In 1961, the bay became the location of an attempted coup against the Cuban government by U.S.-backed revolutionaries. The coup failed and the "Bay of Pigs" became an embarrassment to the United States.

Do things really disappear in the **Bermuda Triangle**?

The "Bermuda Triangle," or "Devil's Triangle," is a popular legend that suggests a supernatural or paranormal reason for a supposedly large number of missing aircraft and sea-going vessels within its area. The legend generally places the area of the Bermuda Triangle in the Atlantic Ocean, with its three corners located at Bermuda, Puerto Rico, and Miami, Florida. But you won't be able to find the Bermuda Triangle on a

Cuba's Communist leader since 1959, Fidel Castro. (Hulton-Deutsch Collection/Corbis)

map since it is not a geographically or politically defined area, and its location is solely designated by the legend.

Though the legend has circulated for at least a century, there seems to be little evidence that this area is subjected to anything but natural hazards and human error. Most of the evidence for the phenomena in the Bermuda Triangle stems from the disappearance of the five aircraft of Flight 19 in December, 1945, as well as a search plane that was sent to find them. Though the popular version of the disappearance of Flight 19 assumes a mysterious end, a mixture of missing navigational apparatus, human error, low fuel, and choppy seas most likely led to the squadron's disappearance and demise.

What was the **first independent country** in the Caribbean?

Haiti was the first independent country in the Caribbean. In 1791, the slaves in Haiti revolted, which led to Haiti's independence from France in 1804. Though Haiti once occupied the entire island of Hispaniola, Haiti now shares the island with the Dominican Republic.

Which Caribbean country leads the region in **tourism**?

The Dominican Republic is the Caribbean country most visited by tourists, attracting one and a half million people annually. The beautiful beaches, clear seas, and tropical climate lure tourists from around the world, but especially from Europe. Jamaica is the second-most visited Caribbean country, with 850,000 tourists annually.

SOUTH AMERICA

PHYSICAL FEATURES AND RESOURCES

Which is **farther east**, Santiago, Chile, or Miami, Florida?

Even though it lies on the west coast of South America, Santiago, Chile, is actually farther east than Miami, Florida. Though it is common to envision South America as directly south of North America, South America actually lies southeast of North America.

Which river carries **more water** than any other in the world?

Though the Amazon River is the second longest in the world (4,000 miles), it carries more water to the ocean than does any other river in the world.

What are the **Andes**?

The Andes are a mountain chain that runs along the entire west coast of South America, from Panama (at the southern tip of Central America) to the Strait of Magellan (at the southern tip of South America). This chain is about 4,500 miles long and contains high plateaus and one of the driest deserts on the planet, the Atacampa. The tallest mountain in South America, Aconcagua (22,834 feet), is located in the southern

What are the Cordillas of Colombia?

In Colombia, the Andes are split into three separate mountain ranges. They are the Cordilla Occidental (western range), the Cordilla Central (central range), and the Cordilla Oriental (eastern range). The city of Cali is located in the valley between the Occidental and Central ranges, while Bogota is located between the Central and Oriental.

Andes, on the border between Chile and Argentina. The ancient Inca city of Machu Picchu is located in the Andes of Peru.

What are the four **climatic regions of the Andes**?

The Andes are known for their four defined climatic zones, which are based on elevation. The lowest zone, tierra caliente (hot lands), is ascribed to the area from the plains to 2,500 feet and is where most of the population resides. The second zone is tierra templada (temperate land), which is from 2,500 to 6,000 feet. The third zone is tierra fria (cold land), which is from 6,000 to 12,000 feet. Above 12,000 feet is the fourth zone, tierra helada (frozen land).

What is the **highest navigable lake** in the world?

Lake Titicaca, located on the border between Peru and Bolivia, is the highest navigable lake in the world, with an elevation of 12,500 feet. Though there are higher lakes in the world, Lake Titicaca is the highest one in which boats can sail. Lake Titicaca was the center of Incan civilization.

Where is the world's **tallest waterfall**?

Angel Falls, in Venezuela, is the world's tallest waterfall at 3212 feet. American pilot Jimmy Angel discovered the waterfall and named it after himself in 1935.

Easter Island. (Archive Photos)

What is the **Atacampa desert**?

One of the world's driest deserts, the Atacampa is located in northern Chile. It is completely barren of plant life. The town of Calama, which is located in the Atacampa, has never received rain. The Atacampa is a source for nitrates and borax.

Who owns **Easter Island**?

Easter Island, which is located east of French Polynesia, is owned by Chile. On this island, there are over 100 large rocks carved into the shape of heads, complete with facial features. These large heads vary in size from 10 to 40 feet and were made out of a soft, volcanic rock.

Is the **Strait of Magellan** crooked?

Yes, it is! The Strait of Magellan is a winding waterway between South America and the islands of Tierra del Fuego at the southern tip of South America. This strait was discovered by the explorer Ferdinand Magellan

207

in 1520 and has been used as a short-cut to avoid having to sail around Cape Horn, the southern tip of South America.

What is the world's largest **tropical rain forest**?

The Amazon forest is the world's largest tropical rain forest. It occupies one-third of Brazil's land area and receives over 80 inches of rain a year. The rain forest loses 15,000 square miles of forest each year because of clear-cutting. The Amazon rain forest is home to about 90 percent of the Earth's animal and plant species and is a major producer of the world's oxygen.

Which country is the world's **leading copper producer**?

Chile produces an astounding 20 percent of the world's copper annually and contains approximately one-fourth of the world's copper reserves. The world's total copper production is 9.2 million metric tons, thus Chilea's production is 1.8 million metric tons.

Which South American country is a member of **OPEC**?

Venezuela, which has about six percent of the world's petroleum reserves, is the only South American country that is a member of the Organization of Petroleum Exporting Countries (OPEC). Venezuela's oil is concentrated in the northern part of the country, in the Maracaibo region.

Which country is the world's **leading coffee producer**?

Contrary to popular belief, it is not Colombia. World coffee production is six million metric tons, and Brazil produces almost 25 percent of that total (Colombia produces only 16 percent). Most Brazilian coffee is grown in the southern part of the country.

Which South American country **exports beef**?

Argentina is the only South American country that exports beef. Argentina has four percent of the world's cattle and exports two percent of the world's beef.

Where does **cocaine** come from?

Cocaine is produced from the coca plant, which was originally domesticated by the Incas. Coca paste from the coca plant is refined to make cocaine. Illegal cartels and individuals in Colombia and other South American countries are major exporters of cocaine, especially to the United States.

HISTORY

Where did the **Incan civilization** develop?

In the 15th and 16th centuries, the civilization of the Incas developed in the altiplanos of the Andes mountains. Altiplanos are high plains located among the mountains that are suitable for habitation. The Bolivian capital of La Paz is also located on an altiplano. The Incan civilization lasted from approximately the 11th through 16th centuries.

What is **Machu Picchu**?

Machu Picchu is an ancient Incan city high in the Andes of Peru. Getting to Machu Pichu requires a twenty-mile hike up the mountains. It was discovered in 1911 by Hiram Bingham, an American, and is believed to have been the last Incan city in the Andes.

How was the **New World divided** between Spain and Portugal?

In 1493 Pope Alexander VI divided the New World into Spanish and Portuguese spheres of influence. A line was placed "100 leagues" (about

How did Brazil become a Portuguese colony?

Most of the present-day country of Brazil was east of the line drawn in the Treaty of Tordesillas, so it became Portuguese territory in 1506. Brazil's official language is Portuguese, making it the only Portuguese-speaking country in South America.

300 miles) west of the Azores islands, located several hundred miles west of Portugal in the Atlantic Ocean. Everything in the New World to the east of this Demarcation Line, which lay off the east coast of South America, belonged to Portugal, while the lands in the west belonged to Spain. Since this division provided little land for Portugal, the Portuguese were dissatisfied. The Treaty of Tordesillas established a new line about 800 miles to the west of the old line. Pope Julius II approved the line in 1506.

Which South American country was the **first to gain independence** from colonial rule?

In 1816, Argentina gained independence from Spain. International recognition of the independent country, then called the United Provinces of the Plate River, did not come until 1823, when the United States recognized the new state.

Which South American territory has **yet to gain its independence**?

French Guyana, on the northeast coast of South America, has been a colony of France since 1817. It is officially a department (state) of France and is the launch site of the European Space Agency.

Ruins of the Incan city of Machu Picchu. (Archive Photos)

What is **Devil's Island**?

Devil's Island, located off the coast of French Guyana, became the overseas prison of France in the middle of the 19th century. France stopped using the island as a penal colony in 1938.

What was **Gran Colombia**?

After many wars against Spain for independence, Gran Colombia, led by Simon Bolivar, became an independent country in 1821. Gran Columbia consisted of the area that is now present-day Colombia, Panama, Ecuador, and Venezuela. In 1830, Gran Colombia was split into Colombia (which included Panama), Ecuador, and Venezuela.

Who was **Simon Bolivar**?

In the early 19th century, Simon Bolivar led the fight in South America for independence from Spain. He is revered as a hero among South

211

Americans for his role in the independence of Venezuela, Colombia, Ecuador, and Peru. Bolivia was named in honor of Bolivar.

PEOPLE, COUNTRIES, AND CITIES

What proportion of South America's population **lives in poverty**?

About one-third of South America's people live in poverty. The continent's wealth is controlled by a small group of people who own most of the land.

Where is South America's **population clustered**?

The population of South America is approximately 329 million (20 percent higher than the United States). In South America, much of the population is found along the Atlantic Coast. Brazil's Atlantic coast is home to two of the world's largest urban areas—Sao Paulo (22 million people) and Rio de Janeiro (10 million people). Buenos Aires, Argentina (11 million people), is also a major urban area on the Atlantic coast.

What are the most **heavily urbanized countries** in South America?

Argentina, Chile, Uruguay, and Venezuela all have an urbanization level (the percent of population who live in urban areas) of approximately 85 percent (the same percentage as the United States).

What is the **largest city** in the Amazon River basin?

Manaus, Brazil, is the largest city in the basin, with a population of just over one million. Manaus is the capital of Brazil's largest state, Amazonas, and is a major trading center for the region. When the Amazon basin was the only known source of rubber, Manaus experienced a boom, but subsequently declined in importance due to the planting of rubber in other regions of the world. Manaus has since recovered by becoming a duty-free trade area.

What is **MERCOSUR**?

The Southern Cone Common Market, also known as MERCOSUR, is a trade group that includes Brazil, Argentina, Paraguay, and Uruguay. It was established in 1995 to reduce trade barriers between those four countries and to promote economic unity.

How large is **Brazil**?

Brazil makes up just under 50 percent of the land area of the entire South American continent. It is the world's fifth-largest country, with 3.3 million square miles of territory.

When did Brazil **move its capital city**?

In 1960, Brazil moved its capital city from Rio de Janeiro to a brand new city in the center of the country, called Brasilia. Brasilia was designed and constructed on empty land near the center of the country in the 1950s. Brazil moved its capital from Rio de Janeiro to Brasilia to assert its independence, exchanging a colonial capital on the coast for a new interior capital. The interior, underdeveloped, location of the new capital allowed a fresh start as well as an opportunity to develop the region.

What statue overlooks **Rio de Janeiro**?

The 100-foot-high statue of "Christ the Redeemer" stands, arms out-stretched, over the city of Rio de Janeiro. The statue of Jesus Christ, with its base on top of Corcovado Mountain at 2340 feet, was built in commemoration of the 100th anniversary of Brazilian independence.

Christ the Redeemer, on the peak of Corcovado, overlooking Rio de Janeiro. (Marilyn Bridges/Corbis)

What are the capitals of Bolivia?

Bolivia has two capitals—La Paz is the administrative capital, while Sucre is the constitutional and judicial capital. Several countries divide national functions between cities.

What country is crossed by both the **equator and a Tropic**?

Brazil is the only country crossed by the equator at 0° and the Tropic of Capricorn 23.5° South.

What South American city has more **Japanese residents** than any city outside of Japan?

Sao Paulo, Brazil, has more Japanese residents than any other city outside of Japan. Well over two million Japanese live in this urban area. Most originally arrived in Brazil as farmers.

What is **Mardi Gras**?

The Catholic festival Mardi Gras literally means "fat Tuesday" in French. Parades, dancing, and carnivals are all part of this pre–Ash Wednesday celebration. It is very popular in Rio de Janeiro, Brazil (where it is know as Carnival), and New Orleans, Louisiana. The Brazilian festival is a significant source of tourism-related income for the country.

What makes **Brazilian automobiles** run?

Over half of Brazilian automobiles use alternatives to petroleum known as gasohol and ethanol. Gasohol is made from sugarcane and ethanol is made from alcohol. The two fuels are much less expensive than petroleum-based gasoline.

215

Who is Fujimori and what has he done for Peru?

Peruvian President Alberto Fujimori has made dramatic and beneficial changes in Peru since his election in 1990. He dismissed the corrupt Peruvian parliament and replaced it with appointees, established a new constitution, captured the leader of the notorious Sendero Luminoso guerrilla group, and has been working on eliminating terrorism.

Where has **one-third of the population of Suriname** emigrated to since 1975?

Suriname was a Dutch colony until it gained independence in 1975. Since 1975, approximately 200,000 of its residents have emigrated to the Netherlands.

What is the world's **highest capital city**?

La Paz, Bolivia, is the world's highest capital city. La Paz is located high in the Andes mountains at an elevation of 12,507 feet (3,700 meters). It was founded in 1548 by Spanish explorers and is now home to approximately 711,000 people.

What **port** does landlocked Bolivia use?

Having no access to the sea itself, Bolivia made an agreement in 1992 with Peru to use its port at Ilo.

What is the actual name of **Bogota**?

Bogota, Colombia, was originally called Santa Fe. More recently, the name became Santa Fe de Bogota. Today, the Colombian capital is known as Bogota for short, and has about six million people in its urban area.

What is a **cartel**?

A cartel is an organization made up of businesses that band together to eliminate competition. In South America, the term usually refers to drug cartels. Two cartels are prominent in Colombia—the Cali and the Medellin. These cartels are powerful, illicit, drug manufacturing and distribution organizations in South America, and many attempts are made at controlling their activities.

Where in South America are you **most likely to be murdered**?

Colombia has a very high murder rate—approximately nine times greater than that of the United States. Murder is the second-leading cause of death (following cancer) for all age groups; it is the leading cause of death for young adult males. The random murders of street urchins and beggars by armed gangs conducting a "cleansing" is quite common.

How **long** is **Chile**?

Chile stretches approximately 2,700 miles along the western coast of South America. At its widest it is only 100 miles across. Chile is a classic example of an elongated country, which makes governing difficult.

Who's fighting over the **Falkland Islands**?

The Falkland Islands (also known as Islas Malvinas), located near the southern tip of South America, have long been a source of conflict between the United Kingdom and Argentina. Though the islands have been occupied by the British since 1833, Argentina has claimed the islands as its own since the 18th century. In 1982, Argentina invaded the islands, but the British regained possession within a matter of weeks. Argentina still claims the Islas Malvinas and is pursing its acquisition through diplomatic channels.

What is the world's **southernmost city**?

Ushuaia, in southern Argentina, is the world's southernmost city. Ushuaia sits on Tierra del Fuego Island, south of the Strait of Magellan.

What is the **Pan-American Highway**?

Begun in the 1930s, the Pan-American Highway is the result of an international effort to create a highway stretching from Fairbanks, Alaska, to Buenos Aires, Argentina. In 1962 a bridge, known as the Bridge of the Americas, was built over the Panama Canal to continue the highway over the canal. A 100-mile stretch of the highway in eastern Panama still remains unfinished.

What is the **primary religion** throughout Latin America?

Due to Spanish and Portuguese colonization, most Latin Americans are Catholic, about 83 percent. Protestants make up about seven percent of the region, and the rest are atheists, nonreligious, animists, or other religions.

What is a **plaza**?

Most Latin American cities have an open public square at the center of the downtown called the plaza. The plaza is used for festivals and ceremonies and is surrounded by a cathedral and shopping areas.

WESTERN EUROPE

PHYSICAL FEATURES
AND RESOURCES

What are the **Alps**?

The Alps are Europe's most famous mountain chain, running east-west for approximately 700 miles. The Alps stretch from southeastern Spain to the Balkans, and include Mont Blanc, the highest point in western Europe.

Where is the **Rock of Gibraltar**?

The Rock of Gibraltar is a limestone mountain located on the Gibraltar peninsula in southern Spain. The city of Gibraltar, located on this same peninsula, is actually a British colony, and is used as a naval air base. This is the perfect location from which to control the Strait of Gibraltar, the small waterway that connects the Mediterranean Sea with the Atlantic Ocean. Spain has continually advocated a claim for this area but has been consistently unable to retrieve this vital piece of land.

On the opposite side of the Strait of Gibraltar, at the northern tip of Morocco, Spain has its own autonomous community, consisting of Ceuta and Melilla, which is also strategically located to control the Strait of Gibraltar.

219

How wide is the **Strait of Gibraltar**?

The strait, which connects the Mediterranean Sea to the Atlantic Ocean between Africa and Spain, is eight miles wide at its narrowest.

What is Iceland's **leading export**?

Over three-quarters of Iceland's exports are fish. The fish industry employs 12 percent of the nations' workforce, and the country is economically vulnerable to fluctuations in world fish prices.

How many **volcanoes** does Iceland have?

Iceland, formed by volcanoes along the Mid-Atlantic Ridge, is home to more than 100 volcanoes. Of these, more than 20 have erupted over the past few centuries.

What are the **seven hills of Rome**?

The city of Rome sits upon seven hills: Capitoline, Quirinal, Viminal, Esquiline, Caelian, Aventine, and Palatine. According to ancient legend, the first settlement in the area, the city of Romulus, was built upon Palatine.

Where is the **Jutland Peninsula**?

The Jutland Peninsula extends north from Germany and is home to the country of Denmark.

Where is the **Black Forest**?

Located in southwestern Germany, the Black Forest is a densely forested, mountainous region that is a popular location for vacationing, with its many health resorts and wilderness trails. The Black Forest is the source of the Danube River and is renowned for its cuckoo clocks.

HISTORY

What is a **Reich**?

The word "reich" literally means "empire" in German. The First Reich is considered to be the Holy Roman Empire from 800 to 1806 C.E. The Second Reich was Germany, united under Otto von Bismarck, from 1871 to 1918. In 1933, Adolf Hitler established the Third Reich, the Nazi regime, which lasted until the defeat of Germany at the end of World War II, in 1945.

What was the **Potsdam Conference**?

At the end of World War II, the United States, United Kingdom, and U.S.S.R. met at Potsdam, Germany, from July 17 to August 2, 1945, for a conference to determine how to control Germany and other eastern territories. The Potsdam Conference divided Germany and Austria into Soviet, French, American, and British zones of control.

What was the **Berlin Wall**?

At the end of World War II, Germany was divided into four zones, each occupied separately by the United States, the United Kingdom, France,

221

Sentries face each other from opposite sides of the Berlin Wall. (Archive Photos)

and the U.S.S.R. The city of Berlin, while located entirely within the Soviet-occupied zone, was itself divided into four zones. Soon thereafter, the Soviets stopped cooperating with the other Allied powers. The three zones occupied by the U.S., U.K., and France joined together to create West Germany, while the Soviet zone became East Germany. A similar split occurred in the city of Berlin.

The city of Berlin held the dichotomy of east versus west, Communist versus capitalist. Many people who lived in East Berlin could see that those in West Berlin generally had a higher standard of living. It is estimated that over two million East Germans fled to the West within Berlin. In August, 1961, the Communist government, determined to stop this mass exodus, began to build the Berlin Wall, a wall that physically divided East and West Berlin. On the west side, the wall became the location of spray-painted messages that voiced free opinions; on the east side of the wall lay a deserted area of barbed wire and armed guards called "No Man's Land."

For decades, the Berlin Wall stood as the physical version of the psychological "iron curtain" that separated east from west. On November 8,

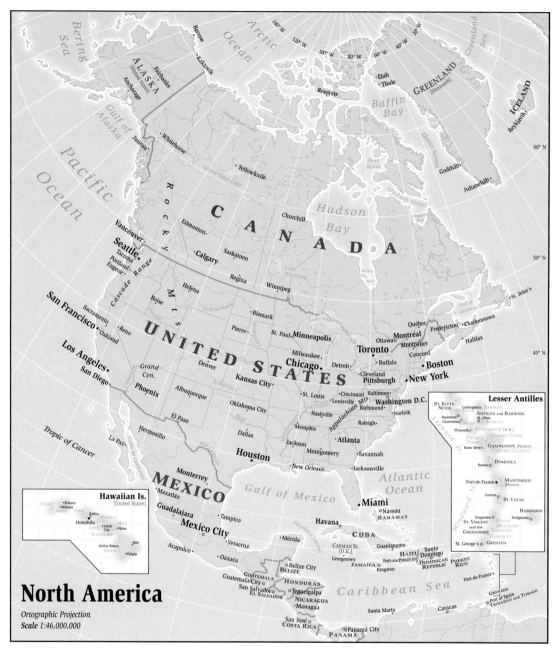

North America

Ortographic Projection
Scale 1:46,000,000

Hawaiian Is.
[United States]

Lesser Antilles

Dreamline Cartography

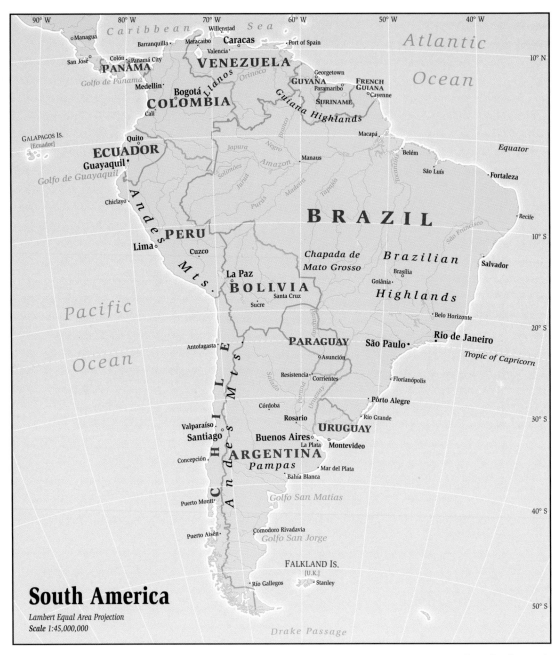

90° W   80° W   70° W   60° W   50° W   40° W

Managua

Caribbean Willemstad *Sea*

Barranquilla Maracaibo **Caracas** Port of Spain

Atlantic

San José Colón Panamá City Valencia 10° N

PANAMA **VENEZUELA** Georgetown *Ocean*

Golfo de Panamá **Llanos** *Orinoco* **GUYANA** Paramaribo **FRENCH GUIANA**

Medellín **Bogotá** **SURINAME** Cayenne

COLOMBIA *Guiana Highlands*

Cali Macapá

GALAPAGOS IS. Quito Belém *Equator*

[Ecuador] *Japurá* *Negro* Manaus São Luís

ECUADOR *Solimões* *Amazon* **Fortaleza**

Guayaquil *Juruá* *Purus* *Madeira* *Tapajós*

Golfo de Guayaquil *São Francisco* Recife

Chiclayo **B R A Z I L**

Andes 10° S

Lima **PERU** Cuzco *Chapada de* *Brazilian* **Salvador**

Mts. *Mato Grosso* Brasília

La Paz Goiânia *Highlands*

BOLIVIA Santa Cruz

Pacific Sucre Belo Horizonte

20° S

PARAGUAY São Paulo **Rio de Janeiro**

Antofagasta Asunción *Tropic of Capricorn*

Ocean Resistencia Corrientes Florianópolis

Salado **CHILE Mts.** Córdoba *Paraná* **Pôrto Alegre**

Valparaíso Rosario Rio Grande 30° S

Santiago **Buenos Aires** **URUGUAY**

Concepción La Plata Montevideo

Andes **ARGENTINA** Mar del Plata

Pampas Bahía Blanca

Puerto Montt *Golfo San Matias* 40° S

Puerto Aisén Comodoro Rivadavia

Golfo San Jorge

FALKLAND IS.

[U.K.]

Río Gallegos Stanley 50° S

Drake Passage

South America

Lambert Equal Area Projection
Scale 1:45,000,000

Dreamline Cartography

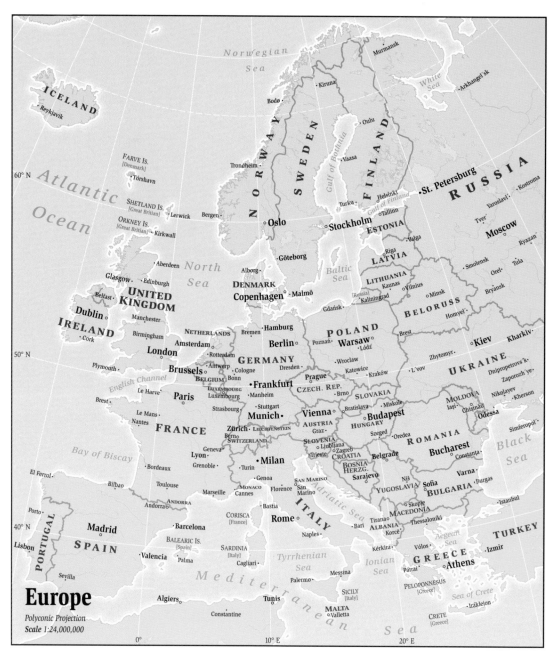

Europe

Polyconic Projection
Scale 1:24,000,000

Dreamline Cartography

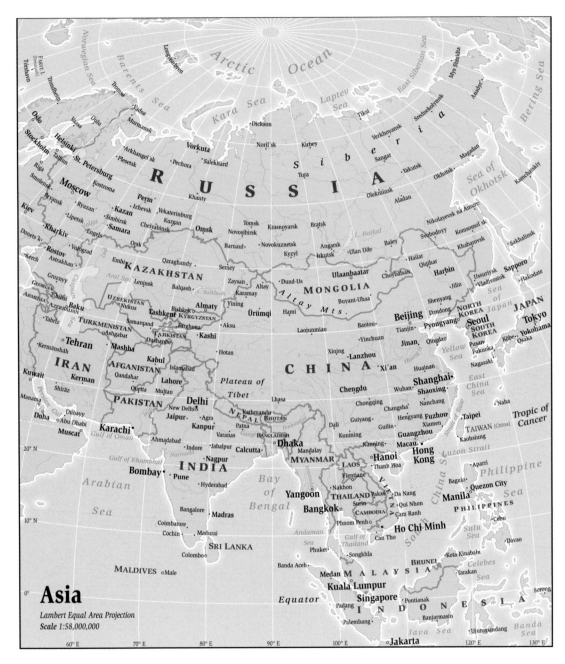

Asia

Lambert Equal Area Projection

Scale 1:58,000,000

Dreamline Cartography

Istanbul

40° E GEORGIA ○Tbilisi 50° E 60° E
Ordu• ARMENIA •Baku 1050° E 60° E
Ankara Yerevan○ AZERBAIJAN 40° N
•Izmîr TURKEY Van• •Tabrīz Caspian Ashgabat○ TURKMENISTAN
Adana• Sea •Mashhad
•Aleppo Mosul• ○Tehran
CYPRUS •Latakia Kermānshāh• Tabas• •Herāt
Nicosia○ LEBANON SYRIA Baghdad IRAN
Mediterranean Beirut• Eşfahān• Kerman
Sea Haifa• Damascus IRAQ •
Tel Aviv• •Amman Basra• •Ābādān 30° N
ISRAEL○ Gaza○ Jerusalem Shīrāz•
Alexandria• Port JORDAN Rafha• Kuwait○ Bushehr• Strait of
Said○ Suez• •Al'Aqabah KUWAIT Hormuz
El Gîza○ Cairo Tabuk• An Nafūd Ad Dammān• •Manama Bandar
Beheshti
•Luxor SAUDI BAHRAIN○ •Manama Gu
Riyadh○ QATAR Doha• •Abu Dhabi
Aswân• ARABIA •Doha U.A.E. •Muscat
Bîr Misahah• Jeddah• Gulf of
Oman
Libya Nubian Rub'al Kha 20° N
Desert Boundary Undefined OMAN Arabian
Desert Dongola• Port Sudan• •Juwara Sea
•Khamīs
Khartoum• Asmara• Sanaa○ YEMEN Gulf of Ade Middle East
SUDAN ERITREA Ta'izz• Mercator Projection
Assab• •Aden Scale 1:52,500,000
DJIBOUTI ○Djibouti

Dreamline Cartography

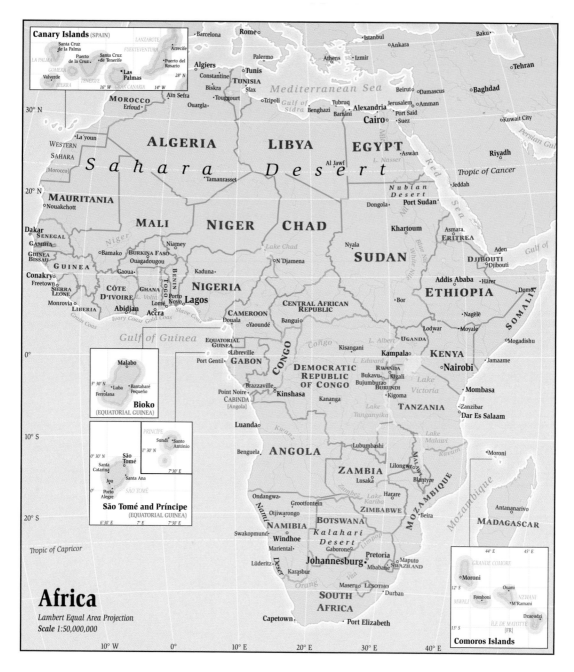

Dreamline Cartography

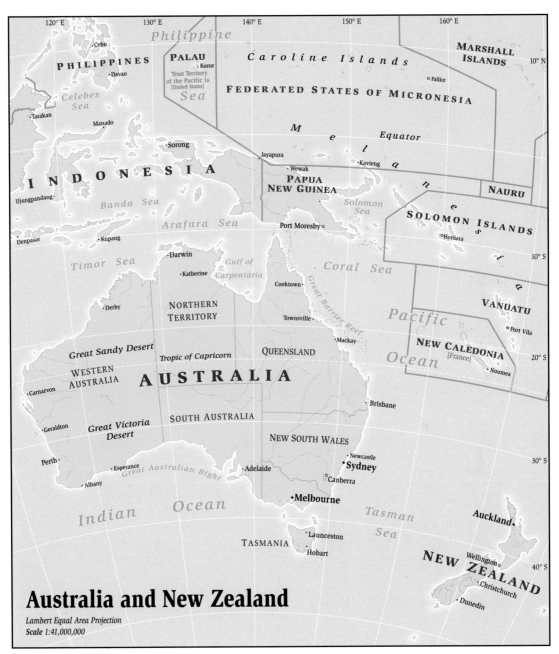

Australia and New Zealand

Lambert Equal Area Projection
Scale 1:41,000,000

Dreamline Cartography

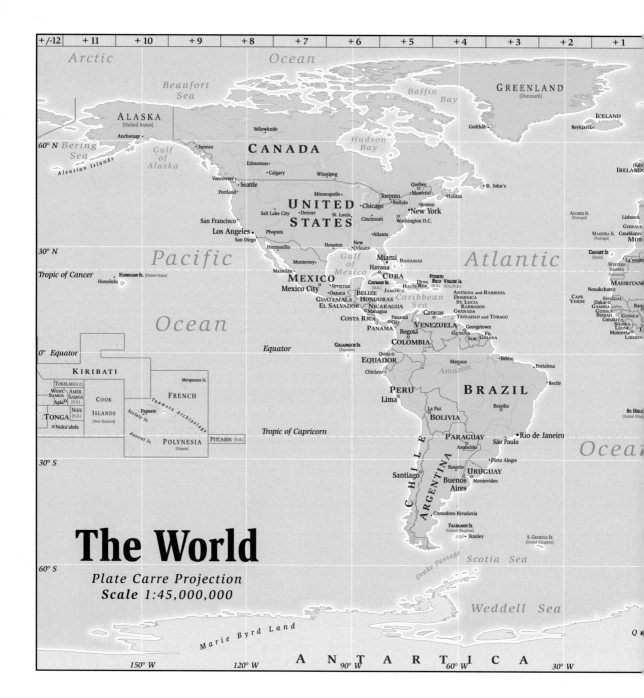

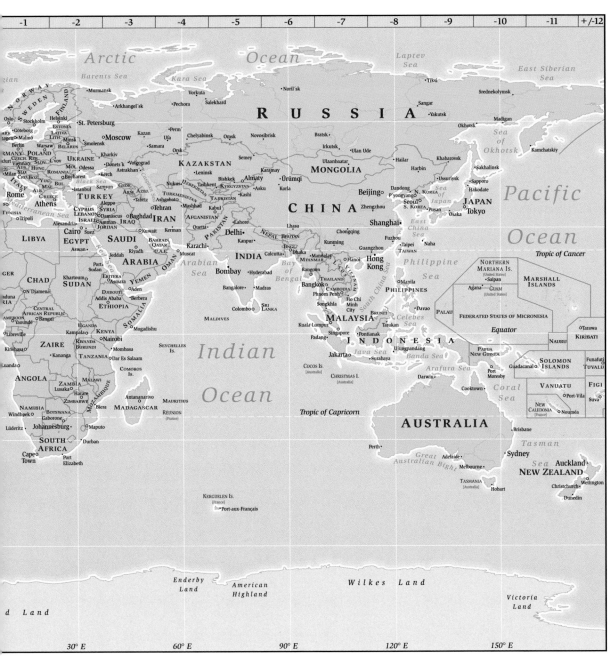

Dreamline Cartography

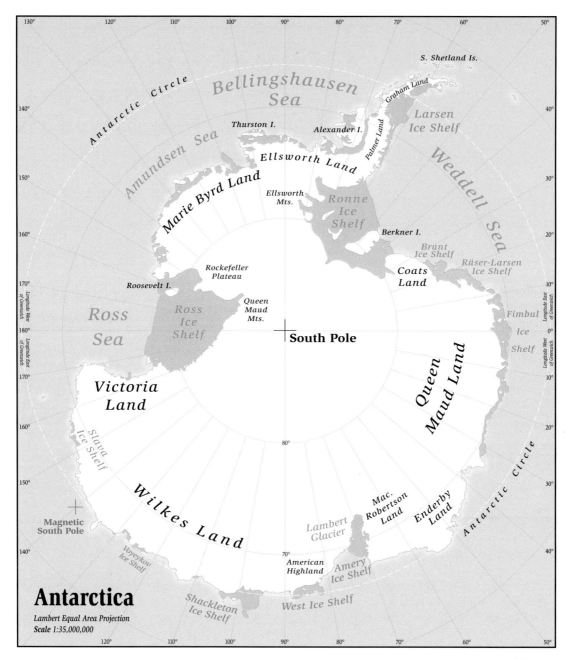

Antarctica

Lambert Equal Area Projection
Scale 1:35,000,000

Dreamline Cartography

What is Machu Picchu? See page 209.
(Archive Photos)

**What is Africa's most populous country?
See page 285.** (Paul Almasy/Corbis)

How many religions have holy sites in Jerusalem? See page 134. (Archive Photos)

Why was the Prime Meridian established at Greenwich?
See page 26. (Dennis di Cicco/Corbis)

What is the world's most visited mountain? See page 256.
(Archive Photos)

What is The Hague? See pages 225 and 226.
(Dave Bartruff/Corbis)

How do cartographers shape our world? See page 19.
(Roger Ressmeyer/Corbis)

What's the big red rock in the middle of Australia? See page 299. (Archive Photos)

What is the lowest point in the world on land?
See page 5. (NASA/Corbis)

What is the longest river in the world?
See page 59. (Jonathan Blair/Corbis)

How many places on Earth have evidence of
asteroid impacts? See page 38.
(Jonathan Blair/Corbis)

Which country has the world's highest population density? See page 132.
(Jonathan Blair/Corbis)

What is the continental divide? See page 191. (Lowell Georgia/Corbis)

1989, the Berlin Wall came tumbling down, and soon thereafter the era of the Cold War also ended.

Where was **Checkpoint Charlie**?

Checkpoint Charlie was a famous crossing point on the Berlin Wall between East and West Berlin, used mainly by tourists and U.S. military personnel.

What is **Hadrian's Wall**?

Hadrian's Wall was built under the direction of the Roman Emperor Hadrian in 122 C.E. Located in northern Great Britain, it was intended to keep out the Caledonians of Scotland. Built of mud and stone, the Wall stretched nearly 75 miles, from Solway Firth in the west to the Tyne River in the east (near Newcastle).

What was the **Maginot Line**?

The Maginot Line was a defensive zone that was built in the 1930s to defend France against the possibility of a German invasion. The zone consisted of underground tunnels, artillery, anti-tank obstacles, and many other defensive structures and stratagems to slow down invading Germans. The Maginot Line stretched for approximately 200 miles near the French-German border.

During World War II, when the Germans did invade France, the Germans bypassed the Maginot Line by storming through neutral Belgium. Thus, the Maginot Line had failed its one great test because it was too short. The Line was also rendered obsolete by the fact that it did not provide defense against the new, modern warfare that included aircraft.

How many Irish left during the **Great Starvation**?

In the mid-19th century, Ireland suffered from the "Great Starvation." From 1845 to 1850, a fungus ravaged the potato crops of Ireland,

What is Benelux?

Benelux stands for Belgium, the Netherlands, and Luxembourg, and represents an economic alliance between the three, that was formed in the 1940s. Since Belgium is primarily industrial and the Netherlands is primarily agricultural, the two countries' economies complement each other, a relationship strengthened by an economic union. Luxembourg, which has a varied economy and is extremely small, has long been closely affiliated with its two larger neighbors, and thus also benefits form the union.

destroying the primary food source of Irish peasants. Though many have called this tragic event the "Great Potato Famine," the mass starvation of the Irish people was caused more by the lack of assistance from the British government than by the famine itself. It is estimated that over one million people died during these catastrophic times, and approximately twice that number left their homeland in an effort to find food and solace.

PEOPLE, COUNTRIES, AND CITIES

What is the **European Union**?

In 1951, six western European countries joined together in the European Coal and Steel Community. As more members joined, the organization grew in scope and soon became an organization that helped mend and meld the economies of Europe. In 1993, the European Community was renamed the European Union (EU), with 15 countries as members: Austria, Belgium, Denmark, Finland, France, Germany, Greece, Ireland, Italy, Luxembourg, the Netherlands, Portugal, Spain,

Sweden, and the United Kingdom. The European Union has a flag, an anthem, and will have a single monetary unit (the "Euro") in 1999.

Where are the **low countries**?

Belgium, the Netherlands, and Luxembourg are known as the low countries because of their low elevation.

How do the **Netherlands** keep getting bigger?

For hundreds of years, the Dutch have been expanding the size of their country by building dikes and draining (and reclaiming) land. These lands, known as polders, have greatly expanded the size of the Netherlands and are now considered one of the seven wonders of the modern world.

What is **Randstad**?

The Randstad is a region of the Netherlands that includes the metropolitan areas of Amsterdam, The Hague, Rotterdam, and Utrech. The urban area of the Randstad holds nearly half the Netherlands' population.

What is **The Hague**?

The Hague is a city on the west coast of the Netherlands with an approximate population of 450,000. The Hague is the home of many international organizations, such as the International Court of Justice.

What are the two **cultural groups** that make up **Belgium**?

The Walloons in southern Belgium, called Walloonia, are descendants of the Celts and speak French. The Flemings in northern Belgium, called Flanders, are descendants of German Franks and speak Flemish, a language similar to Dutch. There is little unity within Belgium, for only 10 percent of Belgians are bilingual.

The Peace Palace, home of the International Court of Justice, in The Hague, Netherlands. (Dave Bartruff/Corbis)

Who **settled Denmark**?

Surprisingly, Denmark was not settled by Europeans from the continent directly to its south, but was settled in the 10th century by Danes from Iceland and the Scandinavian Peninsula.

What's the difference between **England, Great Britain, and the United Kingdom**?

Northeast of France lie two large islands; Great Britain to the east and Ireland to the west. On the island of Great Britain there are three regions: England in the southeast, Wales in the southwest, and Scotland in the north. The other island, Ireland, is divided into two political divisions: the region called Northern Ireland in the north and the country of Ireland in the south. The United Kingdom is a country that includes all three regions on the island of Great Britain (England, Wales, and Scotland) and the one northern region on the island of Ireland (Northern Ireland).

Is **Scotland** a country?

While Scotland does have limited self-rule, it is still part of the country of the United Kingdom. Scotland occupies the northern portion of the island of Great Britain.

What are the **British Isles**?

The British Isles are composed of the two large islands of Great Britain and Ireland (separated by St. George's Channel) and the many small islands nearby. The British Isles include two countries: the United Kingdom and Ireland.

What is the **Commonwealth**?

The Commonwealth, also known as the British Commonwealth, consists of the United Kingdom and the now-independent former countries of the British Empire. The Commonwealth is not a policy-making body but is solely a loose voluntary association between countries that were formerly under British control.

What is **Land's End**?

Land's End has quite an appropriate name, as it is a cape at the southwestern tip of Great Britain that is the westernmost point of England; the "end of land" in the west.

What is a **moor**?

A moor is uncultivated pasture land. You'll find moors in the United Kingdom; in the United States most people call them fields or prairies.

Where is **Camelot**?

The legendary sixth-century castle of King Arthur has been said to be located either near Exeter or Winchester, England. Camelot was not only the home of King Arthur and Queen Guinevere but was also the location of the Round Table and its famous knights.

Did people bathe in **Bath**?

Bath, England, was once the home of a large Roman bath. Though the ancient city lies buried, the modern city of Bath is also renowned for its hot springs, which once warmed the Romans and now offer a relaxing bath in this resort town.

Where is **Catalonia**?

Catalonia is an autonomous region in northeastern Spain. Because Catalonia is home to more than six million Spanish Catalans, who have their own language and culture, many in the region would like the territory to become an independent nation. Spain does not want Catalonia to secede, as Catalonia is responsible for a sizable portion of Spanish economic production. Catalonia's capital is Barcelona, the host city of the 1992 Summer Olympics.

Where is the **French Riviera**?

The French Riviera, also known as Cote d'Azur, is located in southeastern France, near the border with Italy, along the Mediterranean Sea. The French Riviera is a major vacation spot for Europeans, with its mild Mediterranean climate and beautiful scenery. The tiny country of Monaco is located within the French Riviera, and adds to the Riviera's image of luxury with the multitude of casinos and hotels at Monte Carlo.

Where does the **Tour de France** begin and end?

The Tour de France changes its course each year, but the last leg is always along Paris' famous boulevard, the Champs-Elysées. The bicycle race is approximately 2,000 miles long and takes 25 to 30 days to complete.

Where is **Gaul**?

Gaul was an ancient country that included most of modern-day France. Though it began as a Greek colony, Gaul was a Roman province until the fifth century C.E., when the Roman Empire fell. Other empires took control of Gaul, and it eventually became the kingdom of the Franks, the forerunners of the French.

What was the **French Community**?

Only a viable association during the 1950s and 1960s, the French Community, also known as Communaute francaise, was an organization tying together the former French colonies with France.

229

Parliament Square in London, England. (Hulton-Deutsch Collection/Corbis)

Who rules **Andorra**?

Since 1278, the tiny country of Andorra, nestled in the Pyrenees between France and Spain, has been jointly ruled by two people who live outside the country: the President of France and the Bishop of La Seu d'Urgell in northeastern Spain. France and Spain jointly take responsibility for the defense of Andorra.

What are the **largest cities** of Europe?

Western Europe's largest city is Paris, with nine million inhabitants. London (seven million), Milan (four million), Madrid (four million), and Athens (four million) are the next four largest cities.

Which country had the **world's first legislature**?

Though Iceland had been settled but 60 years earlier by the Norwegians, Iceland's parliament, the Althing, was created in 930 C.E.

What is the European country most visited by tourists?

Over 50 million visitors arrive in France annually, making it the most visited European country.

What was the **first tunnel** through the Alps?

The Mont Cenis railroad tunnel was the first tunnel through the Alps and the first major railroad tunnel in the world. Opened in 1871, the tunnel spanned 8.5 miles and connected France and Italy.

Which European country relies the most on **nuclear power** for energy?

France is the European country that relies most on nuclear energy, deriving three quarters of its energy from nuclear power.

What is Europe's **oldest independent State**?

San Marino claims to have been founded in the year 301 C.E. Their first constitution was established in 1600. San Marino is located on Mount Titano in Italy and, at 24 square miles in area, is one of the world's smallest independent countries.

RUSSIA AND EASTERN EUROPE

RUSSIA AND THE FORMER U.S.S.R.

Where does **Asia** end and **Europe** begin?

Though Europe and Asia are actually part of one large landmass, tradition has split the region into two continents along the Ural Mountains in western Russia.

What was the **U.S.S.R.**?

The country called the Union of Soviet Socialist Republics (commonly known as the Soviet Union) was created in 1924, seven years after the Russian Revolution that overthrew the czarist monarchy. The Soviet Union consisted of Russia and its neighboring territories, such as Ukraine, Kazakhstan, and the Baltic States. By the end of 1991, communism had failed in the Soviet Union and many of its internal republics became independent states.

What is the **Commonwealth of Independent States**?

The Commonwealth of Independent States (CIS), established by Russia just after the fall of the Soviet Union, is an organization that serves to keep the resources of the Soviet Union flowing between the now-inde-

Vladimir Lenin, Russia's first Communist leader, in 1917. (Archive Photos)

pendent countries. Twelve of the 15 former Soviet republics are members: Armenia, Azerbaijan, Belarus, Georgia, Kazakhstan, Kyrgyzstan, Moldova, Russia, Tajikistan, Turkmenistan, Ukraine, and Uzbekistan. The three non-members are the Baltic nations of Estonia, Latvia, and Lithuania. The headquarters of this alliance is located in Minsk, Belarus.

How large a part of the Soviet Union was **Russia**?

While the Soviet Union consisted of 15 Soviet Socialist Republics, the largest was the Russian Soviet Federated Socialist Republic (RSFSR). The RSFSR, also known as Russia, comprised three-quarters of the Soviet Union's territory and over half of its population.

How big is **Russia**?

With 6.6 million square miles and 147 million people, Russia is Europe's largest country and also its most populous. Additionally, Russia is the largest country in the world and the world's sixth-most populous. With-

What do the names St. Petersburg, Leningrad, and Petrograd have in common?

St. Petersburg, Leningrad, and Petrograd were three names for the same city. Located in northwestern Russia along the Gulf of Finland, the city was originally founded as St. Petersburg in 1703 by Czar Peter the Great. Since "St. Petersburg" sounded too German to be the capital city of Russia, the city's name was changed to Petrograd in 1914. After the death of Communist leader Vladimir Lenin in 1924, the city's name was again changed, this time to Leningrad. After the fall of the Soviet regime in 1991, Leningrad once again became St. Petersburg.

in Europe, the second largest country is the Ukraine, with 233,100 square miles, and the second-most populous country is Germany, with 82 million people.

What is **Russia's official name**?

Since the fall of the Soviet Union, the official name of Russia has been the Russian Federation. Russia is not just one large political entity, but actually consists of 89 internal regions, provinces, and territories that all have a degree of representation at the federal level.

Why is there a **tiny piece of Russia** in the middle of Eastern Europe?

The important seaport of Kaliningrad, wedged between Poland and Lithuania, was annexed by the U.S.S.R. at the end of World War II. Though once the capital of East Prussia and ethnically German, the Soviets quickly evicted the Germans and replaced them with ethnic Russians. After the Soviet Union collapsed in 1991, many of the autonomous republics within the U.S.S.R. gained independence. Though Kaliningrad

235

How cold is Siberia?

Siberia holds the record for the world's lowest temperature outside of Antarctica. On February 6, 1933, the temperature reached -90 degrees Fahrenheit in Oimyakon, Russia. During the winter, almost all of Siberia has extremely cold temperatures, often reaching -50 degrees Fahrenheit.

lies west of these new countries, its inhabitants are ethnically Russian and thus remained part of the Russian State.

Is it possible to **drive across Russia**?

Since the most prevalent type of road in Russia is a dirt road, driving across Russia is entirely dependent upon the season. During the winter, from November to May, the roads are frozen and can be driven upon; during the summer, many roads become quagmires and are unusable.

How do most people **travel across Russia**?

Most people, as well as goods, travel across Russia by train. In 1891, Czar Alexander III launched the building of a railroad that would unify eastern and western Russia. Traveling from Moscow, through Siberia, to Vladivostok on the Pacific Coast, the Trans-Siberian Railroad was opened in 1904. The Trans-Siberian Railroad is the longest railroad line in the world.

What is the **longest river** in Europe?

The Volga River, which lies entirely within Russia, is Europe's longest river. The Volga River flows 2,290 miles from the Valdai Hills, near the city of Rzhev, into the Caspian Sea.

How big is **Siberia**?

Siberia makes up approximately three-fourths of Russia. Siberia is bounded on the west by the Ural Mountains, on the north by the Arctic Ocean, on the east by the Pacific Ocean, and on the south by China, Mongolia, and Kazakhstan. Russia conquered the area now known as Siberia in the late 16th century.

What is the **largest city in Siberia**?

Located in southern Siberia—and on the Trans-Siberian Railroad route—Novosibirsk is the largest city in Siberia, with a population of 1.4 million.

How big is Russia's **Lake Baikal**?

Lake Baikal, located in southern Siberia, holds one-fifth of the world's non-frozen fresh water. Lake Baikal is also the world's deepest lake, with a maximum depth of just over one mile (5371 feet). The crescent-shaped Lake Baikal is also famous for its crystal-clear water and bountiful plant and animal life.

How did **factories in the U.S.S.R.** end up on the east side of the country?

During World War II, the U.S.S.R. enacted their scorched-earth policy as Germany invaded from the west. The scorched-earth policy involved moving everything they could to the east and burning what they couldn't move. Factories were disassembled, shipped by train to the region near the Ural Mountains, and reassembled to keep Soviet industry working. The Ural Region is still a major manufacturing area for Russia.

What started the fighting in **Chechnya**?

Chechnya was once part of a Soviet republic called Chechen-Ingush. After the fall of the Soviet Union, Chechen-Ingush was divided into two

Where was the Pale of Settlement?

During the 18th and 19th centuries, the Pale of Settlement was an area in which Russia attempted to restrict Jewish settlement. It extended from what is now eastern Poland to Ukraine and Belarus. Within the confines of the Pale, Jews were subjected to anti-Jewish regulations as well as mass killings.

internal republics, Chechnya in the east and Ingushetia in the west. Though the Chechens declared independence in 1992, Russia did not approve, and invaded Chechnya in 1994. The Russians crushed the rebellion, killing thousands of Chechens.

What was the world's **worst nuclear disaster**?

In April, 1986, the Chernobyl nuclear power plant in the Ukraine had a major accident that released radiation into the atmosphere. The protective covering of the nuclear reactor exploded and deadly radiation escaped, immediately killing at least 31 people. The radiation exposure that initially occurred is still killing people through related diseases, and this will continue for many years. More than 100,000 people were evacuated from the region, and deaths due to radiation poisoning continue as radioactive isotopes spread across Europe.

EASTERN EUROPE

What are the **Baltic States**?

The three Baltic States of Estonia, Latvia, and Lithuania are so named because they lie on the Baltic Sea. These three countries became independent after the fall of the Soviet Union in 1991. Poland and Finland, which also lie on the sea, are sometimes also included as Baltic States.

What are the **Balkan States**?

The Balkan States and the Baltic States are two completely different regions. The countries lying on the Balkan Peninsula are commonly referred to as the Balkan States (or the Balkans). The Balkan Peninsula itself lies between the Adriatic Sea (east of Italy) and the Black Sea. The countries on the peninsula include Albania, Bosnia and Herzegovina, Bulgaria, Croatia, Greece, Macedonia, Romania, Serbia and Montenegro, Slovenia, and the portion of Turkey that lies in Europe.

What is **balkanization**?

Due to the fragmentary nature of the countries on the Balkan Peninsula, the term "balkanization" has come to mean any divisive region. For example, the division of the Soviet Union into many countries in 1991 has been referred to as the balkanization of the U.S.S.R.

Does **Yugoslavia** still exist?

In 1991, the republics of Yugoslavia succumbed to internal ethnic pressures and broke apart into five independent countries: Croatia, Bosnia and Herzegovina, Macedonia, Albania, and Serbia and Montenegro. Serbia and Montenegro claim to be the successor country to Yugoslavia and continue to use the name. Though recent cease-fire agreements have brought an artificial calm to the region, civil wars and ethnic cleansing continue, making it one of the world's most unstable areas.

What is **ethnic cleansing**?

Ethnic cleansing is the forced deportation or murder (genocide) of a minority group within a region. Horrible acts of ethnic cleansing have taken place throughout world history. Infamous examples of ethnic cleansing include recent persecutions in the former Yugoslavia, the

239

Combine harvesters at work in a Ukrainian field. (Dean Conger/Corbis)

mass killing of Jews during the Holocaust, and the killing of close to two million Armenians in Turkey during World War I.

How did the **Ukraine** help feed the Soviet Union?

Often called the "breadbasket of the Soviet Union," the Ukraine's rich wheat harvests were used to feed the U.S.S.R. Now, as an independent country, Ukraine produces four percent of the world's wheat and exports much of it to Russia.

Why does **Romania** have so many **orphans**?

The draconian population policies of Romania, which forbade birth control and abortion and required women to have five children, led to the birth of far more children than could be supported by the country. The population policies were implemented by communist dictator Nicolae Ceausescu, who ruled from 1965 until his capture and execution in 1989. The result of his policies has been thousands of Romanian chil-

How unhealthy are Poland's rivers?

Under Communism, industrial pollution went unchecked and Poland's environment has suffered greatly from this negligence. Over 95 percent of the water in Poland's rivers and lakes is considered unfit for consumption. Much of the water has also been considered unfit for agricultural use; two-thirds of the country's water is so contaminated that it corrodes industrial equipment.

dren living in orphanages. Elections were held in 1990 and Romania is now struggling to improve conditions.

Why did **Macedonia's name** cause problems between that country and Greece?

When Macedonia declared independence in 1991, Greece felt indignant that a modern country would use what it felt was a historically Greek name. Greece blocked trade to Macedonia until 1995, when the two countries signed an agreement of understanding. Now, the official name that Macedonia uses in the United Nations is The Former Yugoslav Republic of Macedonia.

Is **Transylvania** a country?

The home of Count Dracula, Transylvania is a region located in central Romania. Transylvania is surrounded by the Transylvanian Alps and the Carpathian Mountains.

Where is **Crimea**?

Protruding into the Black Sea, Crimea is a diamond-shaped peninsula attached—by the Isthmus of Perekop—to southern Ukraine. Though

How has the word bohemian come to mean an unconventional person?

Though Bohemia is a region in the Czech Republic, the term bohemian is sometimes used to refer to an artistic or eccentric person. This term comes from the misguided belief that Gypsies (Roma) originated in Bohemia (it is now known that Gypsies originated in India rather than Bohemia). The stereotypes that are often attributed to Gypsies were then attributed to people from Bohemia, or Bohemians. Now, the word bohemian with a lowercase "b" is used to refer to an unconventional person, while the same word with a capital "B" refers to a person from Bohemia.

Crimea declared its independence from Ukraine in 1992, it later compromised and became an autonomous republic of Ukraine. But this compromise is only a temporary solution, as Crimea is still pursuing its independence.

Where is the **Putrid Sea**?

The Putrid Sea, also known as the Sivash Sea, lies to the east of the Isthmus of Perekop, between Crimea and Ukraine. It is a swampy area of salty lagoons.

What was the **Byzantine Empire**?

After the fall of Rome in 476 C.E. (Common Era), the eastern portion of the Roman Empire became the Byzantine Empire. The capital city of the Byzantine Empire was moved far east to Constantinople (now Istanbul). At one point, the empire included most of the eastern and southern coast of the Mediterranean Sea. The empire shrank in size until 1453, when Constantinople was conquered by the Ottoman Empire.

What two countries emerged from **Czechoslovakia**?

Czechoslovakia was created at the end of World War I (1918) by the Allies as a new country containing the area where the Czechs and Slovaks live. In 1967 and 1968, Czechoslovakia attempted to move away from Communism, but the Soviet Union and Warsaw Pact countries invaded, squelching such aspirations. This "Prague Spring," as it was known, was the first military action taken by the countries of the Warsaw Pact. In 1992, the two republics of Czechoslovakia agreed to divide into two independent countries—the Czech Republic and Slovakia. The dissolution of Czechoslovakia was a peaceful one.

Where is **Atlantis**?

Atlantis, the legendary underwater utopia supposedly located west of the Pillars of Hercules (the land on either side of the Strait of Gibraltar), was first described by Plato in the fourth century B.C.E. (Before the Common Era) as a magnificent civilization that was swallowed by the sea. Though Plato believed Atlantis to have been destroyed, the legend has grown over the centuries to describe this civilization as an underwater kingdom. Researchers now believe that the legend of Atlantis was based on the ancient Minoan civilization that lived on the Greek islands of Thira and Crete, which disappeared after a volcanic eruption in the 16th century B.C.E. Thus, the Minoan civilization in Thira and Crete fits the approximate date of Atlantis' destruction, but not its supposed location.

What two cities make up **Budapest**?

Budapest, Hungary, is actually two cities—Buda and Pest. The two cities are separated by the Danube River; Buda is on the west bank and Pest is on the east bank. The province in which the twin cities are located is also called Budapest.

Who are the **Maygar**?

Maygars are the predominant ethnic group of Hungary. The group originated in Asia, east of the Ural Mountains, and have a language much different from any other in Europe.

What are steppes?

Common throughout Russia, Asia, and central Europe, a steppe is a dry, short-grass plain that can be flat or hilly. While most steppes were once forested areas, cultivation and overgrazing by animals have left only short grasses and barren landscapes.

Do Caucasians come from the **Caucasus Mountains**?

In the latter half of the 19th century, scientists attempted to divide the world's peoples into "races." Each race was defined by the color of its skin, a process that was laced with stereotypes. These scientists used the term "Caucasian" for "white" people because they believed that the region of the Caucasus Mountains of Southwest Asia was their origin. Since that time, we have learned that there is but one human race and that it originated in Africa.

Which **river touches more countries** than any other?

The Danube River, which begins in Germany, passes through or borders 10 countries in Europe, more than any other river in the world. On its journey, the Danube River encounters Germany, Austria, Slovakia, Hungary, Croatia, Serbia and Montenegro, Romania, Bulgaria, Moldova, and Ukraine.

ASIA

Where does **Europe** end and **Asia** begin?

Though Europe and Asia are actually part of one large landmass, tradition has split the region into two continents along the Ural Mountains in western Russia.

CHINA, HONG KONG, AND MIDDLE ASIA

What is the **most populous country** in the world?

China is home to more than 1.2 billion people (1,200,000,000), which means that one in every five people on the planet is Chinese. India, the second-most populous country, has 970 million people and is expected to surpass China's population in the 21st century. Both countries have much larger populations than the third-largest country, the United States, which is home to 270 million people.

What is China's **one-child** rule?

In the late 1970s, the government of China decided that population control was needed because of a rapidly growing population that would soon outgrow the country's ability to feed itself. The policy mandated that every couple would only be allowed to have one child (minorities are exempt from the rule). Punishment is strict and is primarly economic—China won't provide medical care or education funding for the second child. The one-child rule has been working and has slowed population growth, reducing the threat of overpopulation.

What is the **Forbidden City**?

The Forbidden City, located in the center of Beijing, China, was the home of the emperor and the entire imperial court for close to 500 years. Completed in 1420, the Forbidden City (also known as the Purple Forbidden City or Gugong) was the home of 24 emperors from the Ming and Qing dynasties. For centuries, visitors were not allowed into this imperial city. In 1950, several decades after the last emperor was expelled, the city was made a museum and opened to the public.

How was **Taiwan created**?

In 1949, following the Communist revolution in China, the Chinese Nationalist government, led by Chiang Kai-shek, fled to the island of Taiwan and established a Chinese country there. The U.S. and United Nations recognized the government on Taiwan as the rightful government of China and ignored the Communist government on the mainland. In 1971, the (Communist) People's Republic of China replaced Taiwan in the United Nations. Now, the U.S. does not recognize the government of Taiwan.

Is the Great Wall of China the only **man-made object** that can be seen from space?

No, it is not. Besides the Great Wall of China, there are many other man-made structures visible with the naked eye from space, such as urban areas and highways.

PLAN YOUR FAMILY
FOR A BETTER FUTURE

A planned-parenthood billboard in China. (UPI/Corbis)

How **long is China's Great Wall**?

Located in northeastern China, the Great Wall stretches approximately 1,500 miles. It averages 25 feet tall, is 15–30 feet wide at the base, and 10–15 feet wide at the top. The wall was originally erected to keep northern invaders out of China. Though part of the Wall was initially built in the third century B.C.E., the wall was expanded over the course of succeeding centuries.

What is the **underground pottery army**?

In the third century B.C.E., the very first emperor of China, Qin Shi Huang, united China, established the longest-running form of government, built the Great Wall, and built his own elaborate tomb. As a symbol of his rule and to guard himself in the afterlife, Qin Shi Huang had an entire army replicated—approximately 6,000 terra-cotta soldiers and horses, each with distinct facial expressions—to guard the first emperor's mausoleum at Mount Li in Xi'an, China.

The Great Wall of China. (Archive Photos)

248

What is the **Three Gorges Dam**?

The Yangtze River will soon be the site of world's largest electricity-generating facility, the Three Gorges Dam. Construction began in 1994 and should be completed around 2009. The construction of the dam requires the relocation of over one million Chinese living upstream of the dam, as the rising waters will flood towns and archaeological sites. The reservoir will be approximately 600 miles long. The dam will be about 600 feet high, and 1.5 miles wide.

What is the **oldest European settlement** in eastern Asia?

In 1557, the Portuguese established the trading colony of Macao on mainland China at the mouth of the Xi (Pearl) River. Macao has been a territory of Portugal since 1849, but will be returned to China in 1999.

Why isn't **Tibet** on the map?

Tibet is not on the map because it is no longer an independent country. Though it was once a theocratic Buddhist kingdom, China annexed Tibet in 1950. It is now a mildly autonomous region in southwestern China with a puppet communist government. In additon to the destruction of the Tibetan Buddhist religion in the 1960s by China, China also moved Tibetans out of the area and moved ethnic Chinese into Tibet to help moderate Tibet's secessionist ideas.

How long have **Communists been in power** in China?

China's communist revolution took place in 1949, and Mao Zedong became the country's first "chairman." Communism has been the doctrine ever since.

What is the world's most **commonly spoken language**?

Over one billion people around the world speak Mandarin, the official language of China. The world's next most-common language, Hindi, is spoken by only half as many people. Other languages spoken in China

What is pinyin?

Pinyin is a new system for transliterating Chinese into the Roman alphabet. It replaced the Wade-Giles system in 1958, when the Chinese government started using pinyin for external press announcements. It has gradually gained acceptance, and is the reason why we now call the Chinese capital Beijing instead of Peking (a Wade-Giles transliteration).

include Yue (Cantonese), which is spoken by 71 million people; Wu (Shanghaiese), spoken by 70 million; and other minority languages such as Minbei (Fuzhou), Minnan (Hokkien-Taiwanese), Xiang, Gan, and Hakka dialects.

How much **rice** does China produce?

China is the world's leading rice producer, and is responsible for over one-third of the world's rice, about 190 million metric tons. Thailand, however, is the world's leading rice exporter; they ship about a third of the world's rice exports.

What transition took place within **Hong Kong** in 1997?

The control of Hong Kong, a former British territory, was transferred to China on July 1, 1997. Hong Kong, now also known as Xianggang, had been under British rule since 1842. Hong Kong will retain a capitalist economy, but new laws have been created that limit personal freedoms.

What is the **least-densely populated country** in the world?

Mongolia (not to be confused with Inner Mongolia, which is a province in northern China), with its tiny population of two and a half million

> ## Why is the Aral Sea shrinking?
>
> The area of the Aral Sea, located on the border of Kazakhstan and Uzbekistan, has been reduced in size by one-half since 1960. Though it was once the world's fourth-largest lake, diversion of its feeder rivers for agricultural purposes has severely shrunken the lake. This shrinkage has exposed soil saturated with salt, which now destroys plants and vegetation across the nearby plains.

people spread over its 600,000 square miles, has a population density of about four people per square mile. Mongolia's density is limited because only one percent of the country can be used for agriculture; the remainder of the country is dry and used for nomadic herding. Mongolia was originally established in the 13th century when Genghis Khan overtook and unified much of mainland Asia.

Why did the **Soviet Union** invade **Afghanistan**?

In 1979, the Soviet Union invaded Afghanistan in an attempt to install a pro-Soviet government. A civil war ensued, killing two million people. The war lasted a decade, until Soviet troops withdrew in 1989.

Who are the **Sherpa**?

The Sherpa are an indigenous ethnic group in Tibet and Nepal. They live among the mountains of the Himalayas and are often hired as guides for climbing expeditions to such peaks as Mount Everest. In 1953, Tenzing Norgay (Sherpa) and Edmund Hillary (British) were the first two people to reach the 29,028-foot summit of Mount Everest.

THE INDIAN SUBCONTINENT

What is in the **Taj Mahal**?

Located in Agra, India, the Taj Mahal is a mausoleum for the wife of the Mogul emperor Shah Jahan. After Arjuman Banu Bagam's death in 1631, her husband began construction of the mausoleum in 1632. Over 300 feet tall, the white marble mausoleum is a grandiose and striking memorial to her life and death.

What makes **New Delhi** so new?

From 1773 to 1912, the capital of India was Calcutta, then was moved to Delhi in 1912. Because the British wanted to build a brand-new capital city, they began to construct a new city adjacent to Delhi (now known as Old Delhi). When construction was completed on this new city in 1931, it became the capital of India and was known as New Delhi. The metropolitan area of Old and New Delhi is one of the world's largest urban areas, containing a population of 11 million people.

Where did **Bombay** go?

In 1996, India changed the name of the world's fifth-largest metropolitan area (population 17 million) from Bombay to Mumbai.

What is **Bollywood**?

Known as "Bollywood," Mumbai, India (previously called Bombay), is the world's movie capital. The entertainment industry in India produces more films than the United States.

Where is **Dum Dum** airport?

Each year, over two and a half million passengers pass through Dum Dum International Airport in Calcutta, India.

The Taj Mahal, mausoleum for the wife of the Mogul emperor Shah Jahan. (Corbis-Bettmann)

Does a cashmere sweater come from Kashmir?

The Kashmir goats, which make the fine wool known as cashmere, do come from the Kashmir region in India (a disputed area not controlled by any one nation). This region, which is located in northern India bordering on Pakistan, is plagued by violence and unrest. In 1947 the British decided to split the colony of India into two separate countries—one Hindu (India) and one Islamic (Pakistan). The state of Jammu-Kashmir has a mixed population of Hindus and Muslims, which has led to conflict within the region. India and Pakistan have waged three wars over this region, and violence continues today.

Does India have a **population control program** similar to China's?

Though India does not limit births to one child per family, it does have one of the oldest population control programs in the world. The program, begun in the 1950s, encourages the use of birth control and family planning, and the Indian government provides grants to people who undergo sterilization surgeries. Though India's growth rate has not been reduced by the margin that China's has, the population is growing more slowly than it had been prior to the 1950s.

How does India's **caste system** work?

The caste system is the extremely rigid, hierarchical social class system of India. Based on the ancient Hindu text "Law of Manu," the caste system consists of four categories: Brahmans (priests), Kshatriyas (warriors), Vaisyas (merchants), and Sudras (servants). There are many people who are considered to be outside of the caste system, or who have no caste, called the Harijans or untouchables. The untouchables serve the most menial jobs and live literally outside of the social class system. The caste system in India dictates not only one's profession, but also whom one can marry, social contacts, and all other aspects of life.

Where is the world's **second-highest mountain**?

K2, at 28,250 feet, is the world's second-tallest mountain. K2 sits in the disputed Kashmir region of northern Pakistan.

Where was **East Pakistan**?

When the British left South Asia, they divided the region into India and Pakistan. Muslim regions in Pakistan were located to the east and west of Hindu India. These two separate territories became East and West Pakistan, and were separated by approximately 1000 miles. Having been extremely geographically separated from West Pakistan for over three decades, East Pakistan declared independence and became Bangladesh in 1971.

Why does **Bangladesh flood** so often?

Given that most of Bangladesh's elevation is near sea level, and that it is located within the delta of the Ganges and Brahmaputra Rivers, it is not remarkable that the country is easily flooded by regular monsoons (periodic winds and accompanying rainfall) and hurricanes. Unfortunately, Bangladesh also suffers from a poor emergency warning system, so people in Bangladesh are not adequately warned of impending disaster.

JAPAN AND THE KOREAN PENINSULA

JAPAN

What are the **four main islands that make up Japan**?

The northernmost island, Hokkaido, is home to the city of Sapporo. The largest island, Honshu, is the Japanese core area that includes Tokyo and Osaka-Kyoto. The Japanese island of Honshu is the world's seventh-largest island and is the "mainland" island of Japan. It is 86,246 square miles in area, and approximately 100,000,000 people live on the island. In the south are the islands of Shikoku and Kyushu. Kyushu is the southernmost island and was the first island where foreign traders were allowed into Japan. It has a population of over 13 million. Besides the four main islands, Japan includes 2,000 smaller islands. Japan's estimated population for 1998 is 126 million.

What is the world's **most visited mountain**?

Japan's Mount Fuji, a sacred and important volcano to the Japanese, is the country's most popular tourist spot and the world's most visited mountain. Mount Fuji, which is shaped almost like a perfect cone, rises to 12,388 feet and last erupted in 1708.

Where is the **land of the rising sun**?

The Japanese name for Japan, Nippon, which means "origin of the sun," evolved into "land of the rising sun." The name probably derived from the fact that for centuries, Japan was the easternmost known land, and thus where the sun seemed to rise.

How **geologically active is Japan**?

Both volcanoes and earthquakes threaten Japan. Japan has 19 active volcanoes, several of which have erupted in the last decade. Earthquakes are also frequent occurrences, with many very destructive quakes in the last century. In 1923 a major earthquake (approximately 8.3 on the Richter scale) struck Yokohama and killed over 140,000 people. More recently, in 1995, an earthquake in Kobe killed 5,500 people.

How does **Japan get its oil**?

Having no oil itself, Japan must import all the oil it needs. To accommodate the amount of oil necessary, there is a constant stream of oil

Mount Fuji, the world's most visited mountain. (Archive Photos)

tankers, spaced approximately 300 miles apart, that bring oil to Japan 24 hours a day, 365 days a year.

What is the **life expectancy** in Japan?

Japan's life expectancy is the highest in the world. When born, boys have an average life expectancy of 76.7 years, and girls have an average life expectancy of 82.8 years. Japan's infant mortality rate is also the world's lowest (four deaths per 1000 babies born). Japan also has a low natural increase of population growth at 0.2 percent, which means that it will take approximately 350 years for Japan's population to double.

Where is **Iwo Jima**?

Iwo Jima, one of the three islands that make up the Volcano Islands, is located southeast of Japan. One of the deadliest battles of World War II was fought on Iwo Jima, with approximately 20,000 Japanese and 6,000 American soldiers killed. The Japanese air base on Iwo Jima was captured by the United States on February 23, 1945. The island was returned to Japan in 1968.

Where was the **first atomic bomb** used on a populated area?

The Japanese city of Hiroshima was leveled by an atomic bomb dropped by the United States on August 6, 1945. Three days later, the U.S. dropped a second bomb on Nagasaki, Japan. While these events may have hastened the Japanese surrender in World War II, over 115,000 people were killed immediately due to the blasts, and many more died later due to radiation-related diseases.

What is a **bullet train**?

Bullet trains are similar to traditional passenger trains but have been enhanced to travel at speeds of up to 215 miles per hour. Bullet trains have been used in Japan since 1965.

How did **North and South Korea** come to be?

From 1392 to 1910 the Korean peninsula was the home of the Choson Kingdom (as it is referred to by its inhabitants). In 1910 Japan took control of the peninsula but lost this territory after its defeat in World War II. North and South Korea are still divided by a line that lies near the latitude of 38 degrees north. This latitude marked the line dividing the Soviet occupation zone in the north and the American occupation zone in the south following World War II. From 1950 to 1953, the Communist North Koreans fought with the democratic South Koreans. With American forces involved in the Korean war, North Korea's army was eventually forced to retreat into China, though it later recaptured land to the 38th parallel, where the border between the two Koreas remains today.

SOUTHEAST ASIA

Where is **Indochina**?

Indochina is the peninsula in Southeast Asia composed of Myanmar, Thailand, Cambodia, Laos, Vietnam, and the mainland portion of Malaysia. During the colonial era, the eastern portion of Indochina was ruled by France and the west was ruled by Britain.

When did **Burma become Myanmar**?

In 1989, the name of Burma changed to Myanmar when the military took control of the country, following the President's resignation as a result of riots and national turmoil.

Why was the **Vietnam War** fought?

After World War II, the Vietnamese fought the French for their independence. Ultimately, the French left Vietnam divided between a Communist north and non-Communist south. The war in Vietnam that involved American troops lasted from 1962 to 1975 and was fought to keep communism out of South Vietnam. In 1975, North Vietnam conquered South Vietnam and established a unified communist state.

Which Southeast Asian country is one of the **world's richest**?

The tiny nation of Brunei (composed of just 2,226 square miles), located on the island of Borneo, is one of the world's richest countries. Brunei's wealth is based on its oil and gas exports. Sultan and Prime Minister His Majesty Paduka Seri Baginda Sultan Haji Hassanal Bolkiah Mu'izzaddin Waddaulah, who is the ruler of Brunei, is one of the richest men in the world.

Which country has the **lowest maximum elevation** in the world?

The Maldives, located southwest of India and composed of 19 clusters of coral atolls (islands), has a maximum elevation of eight feet. Most of the country lies very close to sea level, making any rise in ocean levels extremely destructive. Formerly under British protection, the Maldives became independent in 1965.

What is an **economic tiger**?

An economic tiger is a term applied to any rapidly developing Asian nation with the power and ability to become an influential, international economic powerhouse. South Korea, Taiwan, and Singapore are considered to be the three economic tigers. Hong Kong was once part of this Pacific Rim group, but, since its merger with China, it can no longer be considered an economic tiger.

THE PHILIPPINES

How many **islands make up the Philippines**?

The Philippines are composed of 7,100 islands. Only 1,000 of these are inhabited and 2,500 still remain unnamed. The islands are divided into three main groups: the Luzon region in the north, the Visayan region in the middle, and the Mindanao and Sulu region in the south.

What is the only Catholic country in Asia?

The Philippines is the only Catholic country in Asia— approximately 83 percent of its population is Catholic. Catholicism was firmly implanted in the Philippines when the land was under Spanish rule, from the 16th through 19th centuries.

When did the **United States control the Philippines**?

The Philippines were a Spanish colony until 1898, when the United States took control of the more than 7,100 islands at the close of the Spanish-American War. The islands remained under American control through 1946 (except for a two-year period of Japanese control during World War II). The Philippines became independent in 1946 and leased land to the United States for military bases until 1992, when the U.S. military presence in the Philippines ended.

INDONESIA

How many **islands make up Indonesia**?

Indonesia is composed of over 13,500 islands. Of these, only 6,000 are inhabited. Indonesia is the world's largest archipelago and was formerly known as the Dutch East Indies. The area had been under the control of the Netherlands since around 1600, but declared its independence in 1945 (after being subjected to Japanese rule during much of WWII).

What is the world's most **densely populated island**?

The Indonesian island of Java is the world's most densely populated island. It has a total population of over 107 million, which, when placed within its 51,000 square miles of land, gives it a density of over 2,100 people per square mile. Indonesia's capital city, Jakarta, is located on the island of Java.

What is Southeast Asia's largest **oil-producing** country?

Indonesia produces about 2.5 percent of the world's petroleum. In 1995, this member of OPEC produced 550 million barrels of oil.

Why isn't **East Timor** independent?

Having been a Portuguese colony for nearly four centuries, East Timor declared independence in 1975, following the withdrawal of the Portuguese. Indonesia subsequently invaded East Timor, killing approximately 100,000 people. This human rights violation and additional massacres in 1991 have resulted in international condemnation. The East Timorese continue to fight for their independence.

MIDDLE EAST

THE LAND AND HISTORY

What is the Middle East **in the middle of**?

At one time, it was common to refer to the Near, Middle, and Far East. Though two of the terms have fallen into disuse, the Middle East is still commonly used. The Near East, at its greatest extent in the 16th century, once referred to the Ottoman Empire, which included Eastern Europe, Western Asia, and Northern Africa. The Middle East referred to the area from Iran to India to Myanmar (formerly Burma). The Far East used to refer to Southeast Asia, China, Japan, and Korea.

Where is the **Middle East today**?

It is generally agreed that the Middle East includes Egypt, Israel, Syria, Lebanon, Jordan, the countries of the Arabian Peninsula (Saudi Arabia, Yemen, Oman, United Arab Emirates, Bahrain, and Qatar), Iraq, Kuwait, Turkey, and Iran. Most regional specialists also include the countries of northern Africa (Morocco, Algeria, Tunisia, and Libya). The new countries of Azerbaijan, Georgia, and Armenia (former Soviet republics) are also often included in the region.

Why is the Dead Sea dead?

Water in the Dead Sea is extremely salty (nine times the saltiness of normal ocean water) and kills nearly all the plant and animal life that flows into the sea from the Jordan River. The Dead Sea lies on the border of Israel and Jordan and is the lowest land elevation on Earth.

What was the **Ottoman Empire**?

The Ottoman Empire began as a tiny state in the 14th century but rapidly expanded through conquest of neighboring states. Its greatest expansion was in the 16th century, when it included southeastern Europe, the Middle East, and North Africa. Due to wars with other European countries in the 17th and 18th centuries, the Ottoman Empire began to decline and became known as the "sick man of Europe." The Empire's successor was Turkey, which became an independent country in 1922.

Is the Middle East a **desert**?

Actually, very little of the region is filled with sand dunes and sand storms. Coastal areas are temperate, and several areas boast very pleasant and moist Mediterranean climates. A lack of water is a problem for agriculture throughout the region, so countries are experimenting with technological solutions such as desalinization plants and drip-agriculture for water conservation.

What is a **Mediterranean climate**?

A Mediterranean climate is a climate similar to the one found along the Mediterranean Sea: hot and dry in the summer and warm and wet in the winter. Areas that are renowned for having Mediterranean climates but are not near the Mediterranean Sea are California and Chile.

What is the Fertile Crescent?

The Fertile Crescent is an area located in a crescent-shaped region between the Persian Gulf in the east and Israel in the west. The development of this region, located along the Tigris and Euphrates Rivers, was begun thousands of years ago and was based on the availability of water from these rivers and the fertile soil they deposited. This area was a center of human civilization and was the location of the ancient Mesopotamian Empire.

Is the **Empty Quarter** empty?

The Empty Quarter, also known as the Rub al Khali, is Saudi Arabia's vast, open desert. While the Empty Quarter is devoid of a large population, it is quite valuable, as it contains the world's largest petroleum reserves.

What commodity makes the **Persian Gulf** so strategically important?

The Persian Gulf is strategically very important because a large share of the world's petroleum is transported through it in oil tankers. At the southeastern end of the Gulf is the Strait of Hormuz, a narrow choke-point that can be (and has been) controlled to prevent ships from sailing in or out of the Gulf.

What was **Mesopotamia**?

This ancient region lay between the Tigris and Euphrates Rivers, from contemporary southern Turkey to the Persian Gulf. Mesopotamia was the home, at one time or another, to such civilizations as Babylonia, Assyria, and Sumer.

The pyramids in Egypt served as tombs for Pharaohs and were built from the 27th to the 10th century B.C.E. (Archive Photos)

What was **Babylonia**?

Babylonia was an ancient country along the Euphrates River, in what is now Iraq. It began in the 21st century B.C.E. and was led by King Hammurabi in the 18th century B.C.E.

When did the **Suez Canal** begin to operate?

After a decade of construction, the Suez Canal, built by the French, opened on November 17, 1869. The 101-mile-long canal cuts through northeastern Egypt, making a passageway for ships to sail between the Mediterranean Sea and the Red Sea, which leads to the Indian Ocean.

Where are **Egypt's pyramids**?

The most famous group of pyramids in Egypt is located near the city of Giza, which is just outside of Cairo. This group includes the Great Pyramid, which is known as one of the Seven Wonders of the Ancient World. The pyramids in Egypt served as tombs for Pharaohs and were built from the 27th to the 10th century B.C.E. Other pyramids are located along the Nile River in southern Egypt and the northern Sudan. Approximately 70 pyramids remain in the region.

RELIGION

Which **religions began** in the Middle East?

Judaism, Christianity, and Islam all have their roots in the Middle East, and the region is filled with sites holy to all three religions. Religious tourists and pilgrims from around the world visit the Middle East in huge numbers.

What is the difference between **Islam and a Muslim**?

Islam is a religion and a Muslim is a person who is a follower of Islam.

Who was **Muhammed**?

The prophet Muhammed was the founder of Islam. He was born in Mecca in 571 and fled to Medina later in life. According to the Islamic religion, he received prophecies from God, which were subsequently written in the Koran, the holy book of Islam. Muhammed's death in 632 C.E. led to the expansion of Islam around Eurasia.

Why do people travel to **Mecca**?

The holy book of Islam, the Koran, requires that all Muslims make a journey, known as a pilgrimage, to the city of Mecca at some time in their lives. Mecca, located in Saudi Arabia, is the holiest city of Islam, as it was the birthplace of Muhammed, the founder of Islam in the sixth century, and contains many important religious sites. Non-Muslims are forbidden to enter Mecca.

Each year, approximately two million Muslims from around the world make a pilgrimage to Mecca. (Archive Photos)

What is the **holiest site** in Mecca?

The most important site in Mecca is the Great Mosque (called the Haram). It is located in the center of the city and houses the sacred Black Stone (Ka'abah). Muslims around the world face the Black Stone during prayer.

How many Muslims make a pilgrimage to Mecca each year?

Each year, approximately two million Muslims from around the world make a pilgrimage to Mecca during the last month of the Islamic calendar. While hundreds of years ago the trip took weeks or months to complete from the vast reaches of the Islamic Empire, today most people from great distances use the modern conveniences of travel and fly to Saudi Arabia.

Who are the **Sunni and Shi'ite** Muslims?

The Sunni and Shi'ite are two sects of Islam. About 85 percent of Muslims are Sunnis, while Shi'ism is especially prevalent in Iran. After

Muhammed's death, two relatives claimed to succeed him. Their followers developed into the two sects.

Who is an **ayatollah**?

The title of ayatollah was originally used for outstanding Shi'ite scholars. But in 1979 Iran went through a period of fundamentalism and replaced the secular shah (the former leader) with the ayatollah, a religious and political leader of the nation. Under the ayatollah's rule, traditional social values were mandated.

What is a **theocracy**?

A theocracy is a country ruled by religion. Iran has the world's largest theocracy, ruled by people who are national leaders as well as religious leaders. While there is a secular president, the division of power between mullahs (a Muslim trained in religous law) and the president is poorly defined.

How far did the **Islamic Empire** spread?

At its widest extent, the Islamic empire included most of northern Africa, Spain and Portugal, the Balkan Peninsula, India, Indonesia, Kazakhstan, and the southern reaches of Russia. Islamic influence also spread east into western China. The areas that were under Islamic rule from the seventh through the 16th century C.E. still keep the religion as a major aspect of their culture.

What is the world's **largest Islamic nation**?

Indonesia, the world's fourth-most populated country, is the world's largest Islamic nation. About 87 percent of Indonesians are Muslims. Islam spread to Indonesia during the Medieval era.

Are all Israelis **Jewish**?

No, they are not. About 80 percent of Israelis are Jewish, and the remaining 20 percent are predominantly Arabs.

CONFLICTS AND NATIONS

How did **Israel** become a country?

After World War II, the British felt over-committed and wanted to reduce the number of colonies in their possession. The future rule of Palestine, one of these British territories, was controversial and a consensus could not be reached. Therefore, the British asked the United Nations to resolve the situation. In November, 1947, the U.N. announced its decision to divide Palestine into two countries—one Jewish and one Arab. The Arabs did not agree with the partition plan and a civil war erupted. The British left the region on May 14, 1948, and Jews immediately declared the creation of the State of Israel.

What is the **Gaza Strip**?

This area of land along the Mediterranean Sea at the border of Israel and Egypt was part of Egypt until it was captured by Israel for a brief period in 1956 and 1957 and then permanently in the 1967 war. The PLO and Israel agreed to Palestinian self-rule in the Strip in 1994.

How has the **West Bank** caused conflict?

The West Bank, which refers to the western bank of the Jordan River, was supposed to become part of an independent Palestine at the same time Israel became a state. However, Arab attacks following the United Nations' 1947 proclamation of Israeli and Palestinian statehood led Israel to take over the West Bank when Israel became an independent state in 1948. Following a 1950 truce, Jordan occupied the West Bank, but Israel retook the land during the 1967 war against its Arab neigh-

Jewish men praying at the Western Wall in Jerusalem. (Archive Photos)

bors. Peace talks in the late 1980s led to an agreement between Israel and the Palestine Liberation Organization (PLO) for limited Palestinian self-rule in the West Bank.

Where are Israeli troops stationed in **Lebanon**?

Israeli troops are stationed in a "security zone" in the southern part of Lebanon, the country immediately to the north of Israel. They have been present since the 1982 invasion of Lebanon by Israel. After a peace treaty was signed, the Israeli army stayed in this zone, which represents a buffer between the two countries.

Which member of the United Nations cannot serve on the **Security Council**?

In order to sit on one of the four rotating spots on the nine-member Security Council, a member of the U.N. must be also a member of one of five regional groups. Israel has not been admitted to any of the five regional groups, so it cannot serve on the Council.

271

What was the Persian Gulf War?

In late 1990, Iraq attacked Kuwait, claiming that it was Iraq's long-lost 19th province. A coalition of countries, led by the United States, fought against Iraq. The land war lasted a mere 100 hours (the air war began January 17, 1991, and ended, along with the land war, on February 28) and Kuwait was liberated from Iraq in February, 1991. Most of Kuwait was destroyed, especially oil facilities, by Iraq's "scorched earth" policy.

Why is **Cyprus divided**?

The island of Cyprus, located in the eastern Mediterranean Sea, became independent from the United Kingdom in 1960. In 1974, a coup occurred that overthrew the President. Turkey invaded the island and succeeded in taking control of its northern half. This became the Turkish Republic of Northern Cyprus, which is not internationally recognized as an independent country. Southern Cyprus remained an independent country—the Republic of Cyprus. Currently, a UN peacekeeping force of 2,000 soldiers monitors the cease-fire line between the two areas.

PEOPLE, COUNTRIES, AND CITIES

What are the **largest cities** in the Middle East?

Cairo, Egypt, is by far the largest city in the Middle East, with over 10 million people. Turkey has Istanbul, an urban area with almost nine million residents. Tehran, the capital of Iran, is home to seven million.

Who lives in the City of the Dead?

There lies on the outskirts of Cairo, Egypt, a very old cemetery filled with mausoleums, memorials, and mosque-shaped tombs and shrines. Because of severe overcrowding in Cairo, squatters now live inside these memorials to the dead, in what is called by many the City of the Dead. Recently, the Egyptian government provided the area with electricity and water. Built as a home for the dead, this cemetery is now becoming a home for the living.

What does the suffix **"stan"** mean?

The suffix "stan" (as in Afghanistan) means nation or land. Therefore, Afghanistan literally means "land of the Afghans."

How many **countries end in the suffix "stan"**?

There are eight. Six are former republics of the Soviet Union: Kazakhstan, Turkestan, Turkmenistan, Kyrgyzstan, Tajikistan, and Uzbekistan. Afghanistan and Pakistan are the two others.

Where is **Asia Minor**?

Asia Minor is the term used for the larger part of Turkey that lies in Asia, east of the Strait of Bosporus. This part of Turkey is also known as Anatolia.

Where is the **Maghreb**?

A region of northwest Africa, the Maghreb is composed of Morocco, Algeria, and Tunisia. Sometimes Libya is also included. The name "Maghreb" remains from the term used by the Islamic empire for this region.

Where in Egypt do **most Egyptians live**?

About 95 percent of Egypt's population lives within 12 miles of the Nile River. Since the rest of Egypt mainly consists of desert, the remaining residents live scattered across the country, primarily near oases or along the coast.

Who are the **Kurds**?

The Kurds are a Middle Eastern people who have no country of their own. They live in southeastern Turkey, northern Iran, northern Iraq, and other nearby countries. Since the Kurds have been persecuted by Iraq, the United Nations established a security zone for them north of 36 degrees north in Iraq. The Kurds hope that one day they will have their own nation-state in the region where they now live.

Which country is composed of **seven sheikdoms**?

The country of United Arab Emirates, located on the Arabian Peninsula, is composed of seven sheikdoms (also known as emirates). The seven emirates were defended by the United Kingdom in the 19th century, but merged to become an independent country in 1971. The seven emirates (Abu Dhabi, Ajman, Al Fujayrah, Ash Sharqah, Dubayy, Ra's a' Khaymah, and Umm al Qaywayn) are presently quite autonomous.

How many people **speak Arabic** in the world?

Approximately 235 million people speak Arabic, making it the world's sixth-most common language. The Koran was written in Arabic, which is common throughout the Middle East. Persian, Turkish, and Hebrew are also spoken in the region.

Which Middle Eastern country has the highest percentage of **urban population**?

Ninety percent of Israel's population lives in urban areas, and its density is third only to those of Bahrain and Lebanon, at 702 people per square mile.

Which Middle Eastern country has the highest per capita **GDP and GNP**?

Oil-rich Kuwait leads the other countries of the Middle East with a per capita Gross Domestic Product of around $17,000, and a Gross National Product of $23,000.

Which Middle Eastern country has the **greatest population**?

Iran is the most populous Middle Eastern country, with about 67 million people, and is the 15th-largest country in the world. Iran is followed by Egypt, with 65 million, and Turkey, with 64 million.

Which country has the most **Azeri** (people from Azerbaijan)?

Surprisingly, over 15 million Azeri live in Iran, considerably more than the six million Azeri who live in Azerbaijan, Iran's tiny neighbor to the north.

Which Middle Eastern country currently has the **longest-ruling leader**?

King Hussein of Jordan has ruled his country continuously since 1952. Born in 1935, Crown Prince Hussein was only 17 when he was named by Parliament to succeed King Talial, his father, who was mentally ill.

Who are the **Bedouin**?

The Bedouins are nomadic tribes that have inhabited areas of the Middle East and northern Africa for centuries. Grazing their goat and camel herds, the Bedouins travel wide distances and frequently cross international borders. Many Middle Eastern countries are attempting to halt the crossings. If these countries are successful, the Bedouin culture would be dramatically changed and perhaps even completely destroyed.

Members of the Bedouin, a nomadic tribe of the Middle East. (Library of Congress/Corbis)

276

Is Saudi Arabia a good place to visit?

Saudi Arabia does not allow foreign tourists—only Muslims making pilgrimages, foreign workers, and business people are allowed into the country. Many other Middle Eastern countries have low numbers of tourists because of fear of terrorism and war.

Which country has a **flag of only a single color** and no design or emblem?

The Libyan flag is green without a design. This North African country is located between Egypt and Algeria and has a population of under six million. Muammar al-Kaddafi continues to rule the country, which was bombed by the United States in 1986 for terrorist involvement.

Where is **Greater Syria**?

Some Syrians, Lebanese, Jordanians, and Palestinians see the boundaries of their countries as artificially imposed by colonial rulers, and hope to see a united Greater Syria composed of the four areas.

What was the **United Arab Republic**?

In 1958, non-neighbors Egypt and Syria united to form the country known as the United Arab Republic. This lasted only until 1961, when Syria decided to become independent. Even after the dissolution, Egypt kept the name United Arab Republic for a decade.

How do you **drive to the island** country of Bahrain?

Bahrain is located 15 miles off the coast of Saudi Arabia has been connected by a four-lane causeway (bridge) to Saudi Arabia since 1986. Saudi Arabia paid for the construction of the causeway.

Where does the **name Saudi Arabia** come from?

The name comes from the ruling Al-Saud family, who have been in control of the country since 1932. Saudi Arabia is the world's only country named after its royal family.

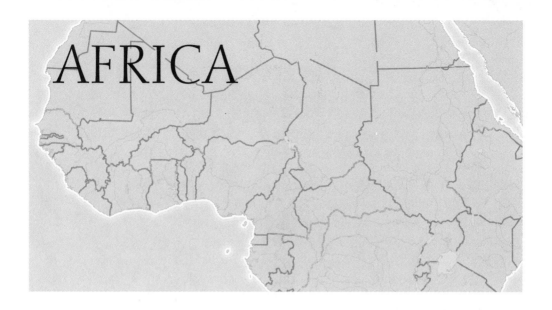

AFRICA

PHYSICAL FEATURES AND RESOURCES

Where is the **Kalahari Desert**?

The Kalahari Desert covers much of Botswana and Namibia. The Kalahari Desert is one of the largest deserts in the world, over 100,000 square miles, and lies on a high plateau at 3000 feet.

How big is the **Sahara desert**?

The Sahara is the world's largest desert. It covers more than 3.5 million square miles in northern Africa. The Sahara receives less than 10 inches of rain each year but contains hundreds of individual oases. The elevation in the Sahara desert ranges from 100 feet below, to more than 11,000 feet above sea level. There are people who live in the Sahara, mostly at or near oases.

Why are the **Blue Nile and White Nile** Rivers both called Niles?

The Nile River begins as two separate rivers—the White Nile and the Blue Nile. The White Nile begins its flow from Lake Victoria in Eastern Africa and the Blue Nile originates in the Ethiopian Highlands. The Blue

Pyramids along the Nile. (Hulton-Deutsch Collection/Corbis)

Nile and the White Nile converge in Khartoum, the capital of the Sudan, and form the Nile River, which continues on to the Mediterranean.

Was the **flooding of the Nile** predictable before dams were built?

The summer floods of the Nile River were so predictable that the Egyptian calendar was based on their rise and fall. Flooding on the Nile occurred from late June until late October. The floods brought nutrients and sediments beneficial to the nearby agricultural lands, making farming productive throughout the remainder of the year. Measuring scales called "nilometers" were placed along the river, and not only measured the river height but also served as a calendars.

Is there any **permanent ice** in Africa?

Though located within three degrees of the equator, there is ice year-round at the top of Mount Kilimanjaro, 19,340 feet in the air.

How was **Mount Kilimanjaro** formed?

Mount Kilimanjaro, the tallest mountain in Africa at 19,340 feet, was formed as a volcano, but is now dormant. The mountain is located in northeastern Tanzania and was first climbed in 1889 by German geographer Hans Meyer and Austrian climber Ludwig Purtscheller.

What is the world's **longest freshwater lake**?

Lake Tanganyika is 420 miles long, the longest freshwater lake in the world, but it's only between 10 and 45 miles wide. It lies on the border between the Democratic Republic of Congo and Tanzania. It is also the world's second-deepest lake, with a maximum depth of 4,710 feet.

What is Africa's **largest lake**?

Located in eastern Africa, Lake Victoria is Africa's largest lake, with a surface area of approximately 27,000 square miles. It is also the world's second-largest freshwater lake, after Lake Superior. Lake Victoria is bordered by Uganda, Kenya, and Tanzania. The lake was named by British explorer John Hanning Speke, the first European to see the lake (in 1858), in honor of the reigning British queen.

What is the **Bight of Bonny**?

Also called the Bight of Biafra, the Bight of Bonny is a bay located in the Gulf of Guinea, near Cameroon. The word "bight" is Old English for bay.

How did a **lake kill** more than 2,000 people?

In August, 1986, Cameroon's Lake Nios, which sits upon a volcanic vent, produced an eruption of carbon dioxide and hydrogen sulfide gasses. The cloud of acidic gas blew into nearby villages, killing more than 2,000 people while they slept.

281

What is the Great Rift Valley?

In eastern Africa there lies a deep valley known as the Great Rift Valley. Over 3,000 miles in length and 20 to 60 miles wide, the rift spans the length of Africa and was formed by tectonic plates sliding apart. In 10 million years, eastern Africa will detach from the rest of Africa along this rift and form its own subcontinent.

What country in Africa has the world's **highest minimum elevation**?

Lesotho, located in the mountains of South Africa, has an absolute minimum elevation of 4,530 feet above sea level in the valley of the Orange River, but almost all of the land in the country lies above 6,000 feet.

Which African country is the world's leading producer of **cocoa beans**?

Cote d'Ivoire, also known as the Ivory Coast, produces approximately one-third of the world's cocoa beans, the beans used to make chocolate. Cote d'Ivoire was named for the large amount of ivory collected from elephant herds that once roamed the country.

Which country produces the **most gold**?

South Africa's gold mines yield 28 percent of the world's gold annually. In 1886, gold was first discovered in South Africa at the mines near Witwatersrand, which is now South Africa's largest gold-producing area.

What **fish**, once thought to be **extinct**, suddenly appeared near Comoros?

In December, 1938, a fisherman discovered a very strange-looking fish near the Comoros islands. Scientists discovered that this lobe-finned fish was a coelacanth, thought to have been extinct for over 70 million years. This species continues to live in the waters off of the Comoros islands, located between Madagascar and continental Africa.

HISTORY

How did the **Berlin Conference** of 1884 help expedite the colonization of Africa?

In 1884, a time when most of Africa had yet to be colonized, 13 European countries and the United States met in Berlin to divide Africa up among themselves. At the conference, geometric borders were drawn across Africa, completely ignoring the continent's cultural differences. Though the borders were established over 100 years ago, they still exist today and are responsible for much of the turmoil and trauma among the now-independent African countries.

How many African countries were **independent in 1950**?

In 1950 there were only four independent countries on the continent: Egypt, South Africa, Ethiopia, and Liberia. All other countries gained their independence in the decades that followed. Most recently, Eritrea became independent of Ethiopia in 1993.

What was **apartheid**?

South Africa's legalized form of racial discrimination was known as apartheid (separateness), and it classified individuals into one of four ethnic groups: white, black, Coloured (mixed race), and Asian. Apartheid laws limited where different groups could live and where they could work. Apartheid was repealed in 1990.

How did American slavery help found Liberia?

In 1822 the American Colonization Society succeeded in founding a colony on the western coast of Africa for freed American slaves. The colony was named Liberia from the Latin word *liber*, which means "free." This colony became independent in July, 1847, and was Africa's first independent country.

Why are there so many **starving people** in Ethiopia?

Droughts in the 1970s and 1980s, which were especially severe from 1984 to 1986, destroyed Ethiopian agriculture. Though relief food was shipped to Ethiopia, internal political corruption kept the food from reaching the starving victims. During the 1980s, approximately one million people died of starvation in Ethiopia.

What was the **Nazis' plan for Madagascar**?

In 1940, the Nazis developed a plan to relocate Jews to the French colony of Madagascar. Once the impracticality of this plan was realized, the Nazis instead decided on an extermination policy toward the Jews.

What is the **Organization of African Unity**?

Founded in 1963, the Organization of African Unity helps strengthen and defend African unity across the continent. The 53 member-countries seek to increase development and economic unity within and between member-countries.

PEOPLE, COUNTRIES, AND CITIES

How many **countries** are there in Africa?

Africa is home to more than a quarter of the world's countries. There are 47 independent countries on the continent itself and 53 if you include the nearby island countries of Comoros, Madagascar, Seychelles, Cape Verde, Mauritius, and Sao Tome and Principe.

How many African countries are **landlocked**?

Only 15 of Africa's 47 continental countries have no access to seas or oceans. These include Botswana, Burkina Faso, Burundi, Central African Republic, Chad, Ethiopia, Lesotho, Malawi, Mali, Niger, Rwanda, Swaziland, Uganda, Zambia, and Zimbabwe.

What is Africa's **most populous** country?

Nigeria, with a population of 107 million, is Africa's most populous country. Nigeria is also the world's tenth-most populous country. If Nigeria's population continues to increase at the present rate, the population will double to 214 million by the year 2022! The average Nigerian woman gives birth to 6.1 children in her lifetime.

How many **provinces** are there in **South Africa**?

In 1994, several new provinces (states) were created. South Africa is now divided into nine provinces: Eastern Cape, Free State, Gauteng, KwaZulu-Natal, Mpumalanga, Northern, Northern Cape, and Western Cape. The nine provinces incorporate homelands of indigenous peoples.

How many **capital cities** does **South Africa** have?

South Africa has three capitals: Pretoria is the administrative capital, Bloemfontein is the capital of the judiciary, and Cape Town is the legislative capital.

Nigerians buying and selling goods on the bank of the Niger in Onishita, Nigeria. (Paul Almasy/Corbis)

Why are so many ships registered in Liberia?

Though most are owned by foreign corporations and spend little time in Liberian territorial waters, many ships register in Liberia because of its low fees for registering ocean vessels. Over 1,600 ships are registered there, giving Liberia one of the world's largest merchant fleets.

How many countries named **Congo** are there?

There are two countries named Congo, and just to make things a little more confusing, they're neighbors. The similarity between the names of the two—the Democratic Republic of the Congo and its western neighbor, the Republic of the Congo—makes distinguishing them a little difficult. In 1908 the Democratic Republic of the Congo was named Belgian Congo; in 1960 it was called Republic of the Congo; in 1964 it became People's Republic of the Congo; in 1966 it was called Democratic Republic of the Congo; in 1971 it was called Zaire; and finally, in 1997, again became the Democratic Republic of the Congo.

Where is **Timbuktu**?

Though the name is commonly used to refer to an extremely distant place, Timbuktu (or Tombouctou) is actually a town near the Niger River, in the African country of Mali. It has a population of about 30,000 and is a major salt-trading post for Saharan Desert camel caravan routes.

Who owns **Walvis Bay**?

When Namibia gained independence from South Africa in 1990, South Africa kept control of Walvis Bay, located approximately 400 miles north of the South African border. The excellent harbor and deep-water port of

287

Walvis Bay was retained for four years, until it was returned to Namibia on March 1, 1994.

Where is the **Barbary Coast**?

The Barbary Coast refers to the countries in Northern Africa that are located along the Mediterranean Sea: Morocco, Algeria, Tunisia, Libya, and Egypt. Though the name Barbary Coast comes from the Berber people who inhabit the region, it is best known for its association with pirates from the 16th through the 19th centuries.

Why is **Cabinda** separate from Angola?

Cabinda, though a province of Angola, is separated from the bulk of Angola by approximately 25 miles of the Democratic Republic of Congo. In 1886, Belgium gave this tiny area to Angola. Cabindans have recently taken up arms against Angola in the hope of obtaining independence.

What was **Zimbabwe's** previous name?

In April, 1980, the British colony of Rhodesia was granted independence and renamed itself Zimbabwe. Rhodesia had been named for a South African businessman, Cecil Rhodes.

Where is **Ouagadougou**?

Ouagadougou (pronounced wah-gah-dah-goo) is the capital of the west African country of Burkina Faso. The city, with half a million residents, is home to the University of Ouagadougou.

Which country lies completely **within South Africa**?

The tiny country of Lesotho, which gained independence from the British in 1966, is completely surrounded by South Africa. Nearly 40 percent of the male workers migrate to South Africa for employment.

Who speaks Swahili?

While approximately 50 million East Africans speak Swahili, it is not an indigenous language. Swahili is a mixture of Arabic and African languages that gradually developed through trading between Africans and Arabs. Though there are over 1,000 different languages in Africa, Swahili is the second-most popular language (Arabic is the first).

Is **Equatorial Guinea** on the equator?

No, Equatorial Guinea's southernmost point is still one degree north of the equator. Though Equatorial Guinea is close to the equator, its southern neighbor, Gabon, is truly equatorial.

Where is the **Horn of Africa**?

The Horn of Africa is the eastern protrusion of Africa that includes Somalia, Ethiopia, and Djibouti. The easternmost tip of the Horn is called Gees Gwardafuy.

How many people lived on the islands of **Seychelles** before 1770?

None. The country, which is composed of 115 islands northeast of Madagascar, was first inhabited by the colonizing French in the 1770s. The British later gained control of the area and brought Africans to the islands. The islands gained independence in 1976.

What is **Caprivi's Finger**?

Caprivi's Finger is the name of the narrow strip of land that protrudes from the northeast corner of the otherwise compact country of Namibia.

289

The land was acquired by German Chancellor Georg Leo von Caprivi from the British in 1890 in order for Namibia, known then as German Southwest Africa, to have access to the Zambezi River. This strip of land is approximately 300 miles long but no more than 65 miles at its widest.

What language do people in **Madagascar speak**?

The people of Madagascar speak Malagasy. Malagasy is closely related to the languages spoken in Indonesia and Polynesia, rather than to any African language. The people of Madagascar are of Indonesian and Malaysian decent, having migrated there nearly 2,000 years ago.

How many **official languages** are there in **South Africa**?

The country has 11 official languages: Afrikaans, English, Ndebele, Pedi, Sotho, Swazi, Tsonga, Tswana, Venda, Xhosa, and Zulu.

Which African country has **Spanish** as an official language?

Equatorial Guinea, which is composed of five islands in the Gulf of Guinea and a tiny sliver of land between Cameroon and Gabon, was a colony of Spain until 1968 and has kept Spanish as its official language. The capital city of Malabo is located on the island of Bioko, previously named Fernando Poo.

Where is the world's **largest church**?

The Basilica of Our Lady of Peace in Yamoussoukro, Ivory Coast, is the world's largest church, covering 100,000 square feet. Built in 1989 by President Felix Houphouet-Boigny, the church has seating for 18,000 people. In 1983, Houphouet-Boigny relocated the capital of the Ivory Coast from Abidjan to his hometown of Yamoussoukro.

How prevalent is AIDS in Africa?

Approximately 14 million Africans have AIDS, which makes up 62 percent of the world's AIDS cases. Almost all AIDS transmission in Africa is through heterosexual intercourse.

Where was **Kunta Kinte** from?

Kunta Kinte, the protagonist of Alex Haley's novel *Roots*, was from Gambia. Though Gambia follows 200 miles of the Gambia River, it is a very thin country, averaging only 12 miles in width. Gambia lies entirely within the west African country of Senegal.

What is the name of the **currency of Botswana**?

The south African country of Botswana, consisting primarily of the Kalahari Desert, uses the pula, which means "rain," as their currency.

What is the only country in the world to provide **constitutional protection to gays, lesbians, and bisexuals**?

South Africa's 1996 constitution protects gays, lesbians, and bisexuals against discrimination in both the public and private sectors. The "Equality Clause" in the Bill of Rights protects people from discrimination based on race, gender, pregnancy, marital status, ethnic or social origin, color, sexual orientation, age, disability, religion, conscience, belief, culture, language, and birth.

Who controls **Western Sahara**?

Located on the west coast of Africa, Western Sahara was a colony of Spain until 1976. Upon the withdrawal of the Spanish, the territory was divided between its neighboring countries, Mauritania in the south and Morocco in the north. In 1979 Mauritania gave up its claim and Morocco took control over the entire territory. The United Nations would like the Western Saharans to vote for independence, but Morocco continually delays such a vote.

OCEANIA AND ANTARCTICA

OCEANIA

What is **Oceania**?

Oceania is the region in the central and southern Pacific that includes Australia, New Zealand, Papua New Guinea, and the islands that compose Polynesia, Melanesia, and Micronesia.

Who **owns all of those islands** in the Pacific Ocean?

There are hundreds of islands in the Pacific Ocean. While many are part of independent countries, others remain colonies or vestiges of colonial empires.

What is **Micronesia**?

The region known as Micronesia consists of islands east of the Philippines, west of the International Date Line, north of the equator, and south of the Tropic of Cancer. Approximately 600 of the islands in this area united in 1986 to form an independent country, the Federated States of Micronesia. Though 600 islands joined the Federation, there are approximately 1,500 other islands in the region, including the independent countries of the Marshall Islands, Kiribati, and Palau.

A mushroom cloud off the coast of Bikini Island. (The National Archives/Corbis)

What is **Polynesia**?

Polynesia consists of islands in the region bounded by Hawaii in the north and New Zealand in the southwest. The region includes the countries of Samoa, Tonga, and Tuvalu. Also located in this area is the colony of French Polynesia, which includes Tahiti and 117 other islands and atolls.

What is **Melanesia**?

Located northeast of Australia, Melanesia is a small region that lies south of the equator and west of 180 degrees longitude, but excludes New Zealand. Melanesia includes the countries of Vanuatu and Fiji as well as the Solomon Islands and New Caledonia.

What is a **coral reef**?

Coral reefs are formed by the accumulation of calcium carbonate that comes from the external skeletons of tiny animals called coral polyps.

The polyps live in shallow, warm water, and thus congregate around islands in the tropics, where coral reefs are abundant.

Where is the **largest coral reef** in the world?

The Great Barrier Reef is the largest coral reef in the world. Located just off the northeastern coast of Australia, it extends for over 1,200 miles. Much of the area is now protected as a marine park.

What is an **atoll**?

In addition to reefs, coral can also form atolls. Atolls are formed when a volcano, around which coral often grows, erodes away, leaving a circular wall of coral with a lagoon at the center.

How did the **bikini** get its name?

In 1946, the United States began to test atomic weapons on the Bikini Atoll in the Marshall Islands. It was also in the late 1940s when the two-piece bathing suit made its debut and took its name from the intensely publicized Bikini Atoll.

How do the people on the tiny islands in the Southern Pacific Ocean go **shopping**?

Most basic goods can be bought on each populated island, and larger items can be shipped via airplane. When residents need to travel between islands, they often take to the air. Each island has an airstrip of adequate length for its own transportation needs. In the past, inhabitants used boats as their primary means of transportation between islands.

Where did **Gauguin** live?

The French painter Paul Gauguin moved to Tahiti in 1891 and later to other islands in French Polynesia to escape European civilization.

How did the Guano Island Act help fertilize America?

In 1856, the United States Congress passed the Guano Island Act, which allowed the U.S. to take possession of any unclaimed island that contained guano. Guano, the excrement of sea birds, was mined for use as fertilizer before the widespread use of chemical fertilizers. Beginning in 1857, the Baker and Howland Islands, located southwest of Hawaii, were mined by the U.S. until their guano was depleted in 1891.

Which country has **more languages** than any other in the world?

Papua New Guinea is home to more than 700 different languages. The most common languages spoken there are Motu and pidgin English.

Where did the **mutineers of the Bounty** land?

In 1789, members of the crew of the HMS *Bounty* mutinied. After having dropped off 19 other members of the crew, including Captain William Bligh, the mutineers landed on the uninhabited island of Pitcairn. While the captain and his loyal crew successfully returned to England, the mutineers established a community composed of nine male mutineers, six male Polynesians, and 12 female Polynesians who had also been on board the *Bounty*. In 1856, approximately 200 of the mutineers' descendants voluntarily moved from Pitcairn Island to Norfolk Island because of overpopulation.

Where did **Charles Darwin** develop his theory of natural selection?

Charles Darwin graduated from Cambridge University in 1831 and spent the next five years of his life as a naturalist on board the HMS *Beagle*. The *Beagle* traveled around the world, including among its destinations the Galapagos Islands, west of South America, where Darwin spent six weeks collecting data from which he developed his theories of natural selection, published in *Origin of the Species* in 1859.

Australia

How did Australia get its **name**?

Long throughout history there remained an assumed, yet completely undiscovered, land called Terra Australis Incognita, or unknown southern land. As early as the fourth century B.C.E. Aristotle believed that an extremely large continent, located in the Southern Hemisphere, lay undiscovered and would complete the symmetry of the land masses. For centuries, this unknown landmass remained a treasured legend and often appeared on maps in varied sizes and shapes. When the territory now known as Australia was discovered in the early 17th century, no one believed that this was the famed Terra Australis Incognita. During the early 17th century, the western coast of this territory was named New Holland and claimed for the Netherlands; in 1770 James Cook claimed the east coast of this territory for England and called it New South Wales. It wasn't until 1803 that Matthew Flinders circumnavigated this territory and proved that it was a continent and was the long sought-after Terra Australis Incognita. Finally, in the early 19th century, nearly two centuries after having been discovered, this land was finally named Australia.

Who are the **Aborigines**?

The Aborigines are the indigenous inhabitants of Australia, having migrated from Southeast Asia approximately 40,000 years ago. In the late 18th century, when European colonization began, there were over 300,000 Aborigines in Australia. Many were killed by European diseases and abuses, and by 1920 there were only 60,000 Aborigines remaining. Like the Maori of New Zealand, Australia's Aborigine population rebounded in the late 20th century, and now stands at over 200,000. Most Aborigines now live in urban areas and are gaining political support and benefits.

Australia's Uluru, or Ayers Rock. (Archive Photos)

Was Australia really used as a **penal colony**?

Yes, approximately two-thirds of Australia's initial settlers were convicts from Great Britain. From 1788 to 1850, when Australia was used as a penal colony, approximately 160,000 prisoners were sent to the continent. Though Britain stopped sending prisoners to Australia in 1850, free colonists began arriving with the first ship of convicts.

Where is the **outback**?

The outback is the general term used to describe the remote interior of Australia. Most of Australia's population is concentrated on the coast, since the interior is extremely dry and barren.

What is the **capital of Australia**?

The capital of Australia is Canberra, which is located in a federal territory (similarly to Washington, D.C.) within the Australian state of New South Wales. When Australia was founded in 1901, both Sydney and

Melbourne wanted to become the capital city. In 1908 it was ultimately decided that a brand-new capital city would be built and located away from the coast. Canberra is the largest city in Australia that is not on the coastline (population 308,000).

What's the **big red rock** in the middle of Australia?

The big, red, sandstone monolith in the center of Australia is called Uluru (its indigenous name; it was previously called Ayers Rock). It is the world's largest monolith, approximately 1.5 miles wide and one-fifth of a mile (1,100 feet) high.

Are **kangaroos** native to Australia?

Yes, kangaroos are native to Australia. Kangaroos range in size from giant kangaroos (five feet tall) to tiny, rat-sized kangaroos, called potoroos.

Is the **Tasmanian devil** a real animal?

The Tasmanian devil is a real animal, though it resembles few characteristics of its cartoon counterpart. The real Tasmanian devil is a carnivorous marsupial that lives on the island of Tasmania, just southeast of the Australian mainland.

Is Australia the **smallest continent**?

Despite being the sixth-largest *country* in the world, Australia is the smallest continent in the world. Australia is approximately three million square miles in area, just a little smaller than Brazil.

Which country is the world's leading **bauxite** producer?

Australia mines more bauxite than any other country in the world, producing 40 percent of the world's supply annually. An ore of aluminum, bauxite is especially prevalent in Australia's Darling Range, located in the southwestern part of the country.

Which country is the world's leading **lead** producer?

Australia produces approximately 17 percent of the world's lead annually. Most of the lead is mined near Mount Isa in the northeast area of the country and near Broken Hill in the southeast.

How much **beef** does Australia export?

Australia is responsible for more than 15 percent of the world's beef exports—739,000 metric tons.

What is a **boomerang**?

The boomerang was developed as a hunting tool by the Aborigines of Australia. There are two types of boomerangs—those that return and those that do not return. The returning boomerang is used to kill small animals; the non-returning boomerang is used to kill large game or enemies.

NEW ZEALAND

Who are the **Maori**?

The Maori are the indigenous inhabitants of New Zealand. Around the ninth century, the Maori arrived in New Zealand from other Pacific Islands. In 1769 there were over 100,000 Maori, but their population decreased significantly (to 40,000) by the end of the 19th century, due to European colonization. In the 20th century, the Maori population has expanded (to 400,000) and Maoris now compose approximately one-ninth of the total New Zealand population.

What is the Royal Flying Doctor Service?

Created in 1928, the Royal Flying Doctor Service (RFDS) is a charity organization established to provide health care and emergency services to the sparse population of Australia's outback. With 38 aircraft, the RFDS averages over 80 flights a day, helping more than 135,000 people per year.

Are there **more people or sheep** in New Zealand?

New Zealanders are outnumbered 16 to one by sheep! There are just over three and a half million people in New Zealand, but nearly 56 million sheep. New Zealand has long been a leading exporter of wool.

Who are **Kiwis**?

One nickname for a New Zealander is "Kiwi," but kiwis are also a flightless bird and a type of fruit found in New Zealand. Kiwi birds, a national icon of New Zealand, have long thin beaks and lay eggs larger in proportion to their body size than any other bird. Most of the world's supply of kiwi fruit is also grown in New Zealand.

What was the *Rainbow Warrior*?

In 1985, the Greenpeace ship *Rainbow Warrior* was in the Auckland, New Zealand, harbor when it exploded and sank, killing one Greenpeace staff member. It was later discovered that French secret agents planted bombs onboard the *Rainbow Warrior* in order to stop the organization from protesting French nuclear weapon tests in the Pacific. Following the incident, the French minister of defense and head of the secret service were forced to resign. New Zealand, a country very much opposed to nuclear weapons, maintained a poor relationship with France for many years following the bombing.

The Antarctic Treaty was established in 1959, proclaiming that no additional claims could be made upon Antarctica and that the continent would solely be used for scientific purposes. (Chris Rainier/Corbis)

Which country was the world's **first welfare state**?

In 1936, New Zealand became the world's first welfare state by offering its citizens full social security and health benefits.

Which country first granted **women the right to vote**?

In 1893, New Zealand became the first country to give women the right to vote.

ANTARCTICA

How **thick is Antarctica's ice**?

Most of the ice in Antarctica is approximately one mile thick. Over 80 percent of the world's freshwater is stored in ice in Antarctica. Some

What is ANZUS?

In 1951, Australia, New Zealand, and the United States signed the Australia–New Zealand–United States (ANZUS) Treaty to protect each other militarily. In 1986, New Zealand banned nuclear weapons from its country and thereafter refused to allow U.S. nuclear-powered or nuclear-armed ships to dock in its harbors. New Zealand was summarily excluded from the Treaty.

have suggested that large chunks of ice be cut off from Antarctica and shipped to dry regions of the world, but this has yet to be done.

Which continent has the **highest average elevation**?

The average elevation of Antarctica, approximately 8,000 feet, is higher than that of any other continent. The highest point in Antarctica is Vinson Massif, with an elevation of 16, 860 feet.

How **dry** is Antarctica?

Though Antarctica is covered with ice, it is the driest continent on the planet. The ice in Antarctica has been there for thousands of years, and the continent receives less than two inches of precipitation annually— the Sahara Desert receives 10 inches each year.

Who **owns Antarctica**?

Though Antarctica is a cold, icy, barren territory, seven countries claimed portions of it in the early 20th century. All of these claims were defined by lines of longitude, and problems arose as many of these claims overlapped. In 1959, the Antarctic Treaty was established, proclaiming that no additional claims could be made upon Antarctica and that the continent would solely be used for scientific purposes.

303

COUNTRIES OF THE WORLD

There are currently 192 countries in the world, and in this chapter you'll find key statistics on each one. What is the fertility rate in Yemen? Which religions are practiced in Vietnam? What is the population of India? What is the total area of Canada? You can use this chapter to find important geographic, political, and cultural information on countries mentioned previously in the book. Use the color maps in the center of the book as a visual guide to the countries listed below.

Afghanistan

Long Name: Islamic State of Afghanistan

Location: Southern Asia, north and west of Pakistan, east of Iran

Capital: Kabul

Government Type: transitional government

Currency: afghani (AF)

Area: 249,935 sq. mi. (647,500 sq. km)

Population: 24,792,000

Population Growth Rate: 4.2%

Birth Rate (per 1,000): 42

Death Rate (per 1,000): 17

Life Expectancy (from birth): 46.8

Total Fertility Rate (average children per woman): 6.0

Religion: Sunni Muslim 84%, Shi'a Muslim 15%, other 1%

Languages: Pashtu 35%, Afghan Persian (Dari) 50%, Turkic languages (primarily Uzbek and Turkmen) 11%, 30 minor languages (primarily Balochi and Pashai) 4%, much bilingualism

Literacy: 31.5%

GDP Per Capita: $800

Occupations: agriculture and animal husbandry 67.8%, industry 10.2%, construction 6.3%, commerce 5%, services and other 10.7% (1980 est.)

Climate: arid to semiarid; cold winters and hot summers

Terrain: mostly rugged mountains; plains in north and southwest

Albania

Long Name: Republic of Albania

Location: Southeastern Europe, bordering the Adriatic Sea and Ionian Sea, between Greece and Serbia and Montenegro

Capital: Tirane

Government Type: emerging democracy

Currency: lek

Area: 10,576 sq. mi. (27,400 sq. km)

Population: 3,331,000

Population Growth Rate: 1%

Birth Rate (per 1,000): 21

Death Rate (per 1,000): 7

Life Expectancy (from birth): 68.6

Total Fertility Rate (average children per woman): 2.6

Religion: Muslim 70%, Albanian Orthodox 20%, Roman Catholic 10%

Languages: Albanian, Greek

Literacy: 72%

GDP Per Capita: $1,290

Occupations: agriculture 49.5%, private sector 22.2%, state sector 28.3%

Climate: mild temperate; cool, cloudy, wet winters; hot, clear, dry summers; interior is cooler and wetter

Terrain: mostly mountains and hills; small plains along coast

Algeria

Long Name: Democratic and Popular Republic of Algeria

Location: northern Africa, bordering the Mediterranean Sea, between Morocco and Tunisia

Capital: Algiers

Government Type: republic

Currency: Algerian dinar (DA)

Area: 919,351 sq. mi. (2,381,740 sq. km)

Population: 30,481,000

Population Growth Rate: 2.1%

Birth Rate (per 1,000): 28

Death Rate (per 1,000): 6

Life Expectancy (from birth): 68.9

Total Fertility Rate (average children per woman): 3.4

Religion: Sunni Muslim (state religion) 99%, Christian and Jewish 1%

Languages: Arabic, French, Berber dialects

Literacy: 61.6%

GDP Per Capita: $4,000

Occupations: government 29.5%, agriculture 22%, construction and public works 16.2%, industry 13.6%, commerce and services 13.5%, transportation and communication 5.2% (1989)

Climate: arid to semiarid; mild, wet winters with hot, dry summers along coast; drier with cold winters and hot summers on high plateau; sirocco is a hot, dust/sand-laden wind especially common in summer

Terrain: mostly high plateau and desert; some mountains; narrow, discontinuous coastal plain

Andorra

Long Name: Principality of Andorra

Location: Southwestern Europe, between France and Spain

Capital: Andorra la Vella

Government Type: parliamentary democracy

Currency: French franc and Spanish peseta

Area: 174 sq. mi. (450 sq. km)

Population: 65,939

Population Growth Rate: 0.5%

Birth Rate (per 1,000): 10

Death Rate (per 1,000): 5

Life Expectancy (from birth): 83.5

Total Fertility Rate (average children per woman): 1.2

Religion: Roman Catholic (predominant)

Languages: Catalan (official), French, Castilian

Literacy: 100%

GDP Per Capita: $18,000

Occupations: agriculture, tourism

Climate: temperate; snowy, cold winters and warm, dry summers

Terrain: rugged mountains dissected by narrow valleys

Angola

Long Name: Republic of Angola

Location: Southern Africa, bordering the South Atlantic Ocean, between
Namibia and Democratic Republic of the Congo

Capital: Luanda

Government Type: transitional government, nominally a multiparty democracy with a
strong presidential system

Currency: new kwanza (NKz)

Area: 481,226 sq. mi. (1,246,700 sq. km)

Population: 11,178,000

Population Growth Rate: 2.8%

Birth Rate (per 1,000): 44

Death Rate (per 1,000): 17

Life Expectancy (from birth): 47.9

Total Fertility Rate (average children per woman): 6.2

Religion: indigenous beliefs 47%, Roman Catholic 38%, Protestant 15%

Languages: Portuguese (official), Bantu and other African languages

Literacy: 42%

GDP Per Capita: $800

Occupations: agriculture 85%, industry 15% (1985 est.)

Climate: semiarid in south and along coast to Luanda; north has cool, dry season (May to
October) and hot, rainy season (November to April)

Terrain: narrow coastal plain rises abruptly to vast interior plateau

Antigua and Barbuda

Long Name: Antigua and Barbuda

Location: Caribbean, islands between the Caribbean Sea and the North Atlantic Ocean, east-southeast of Puerto Rico

Capital: Saint John's

Government Type: parliamentary democracy

Currency: EC dollar (EC$)

Area: 170 sq. mi. (440 sq. km)

Population: 64,006

Population Growth Rate: 0.4%

Birth Rate (per 1,000): 17

Death Rate (per 1,000): 6

Life Expectancy (from birth): 71.2

Total Fertility Rate (average children per woman): 1.7

Religion: Anglican (predominant), other Protestant sects, some Roman Catholic

Languages: English (official), local dialects

Literacy: 89%

GDP Per Capita: $6,800

Occupations: commerce and services 82%, agriculture 11%, industry 7% (1983)

Climate: tropical marine; little seasonal temperature variation

Terrain: mostly low-lying limestone and coral islands with some higher volcanic areas

Argentina

Long Name: Argentine Republic

Location: Southern South America, bordering the South Atlantic Ocean, between Chile and Uruguay

Capital: Buenos Aires

Government Type: republic

Currency: nuevo peso argentino

Area: 1,056,362 sq. mi. (2,736,690 sq. km)

Population: 36,738,000

Population Growth Rate: 1.3%

Birth Rate (per 1,000): 20

Death Rate (per 1,000): 8

Life Expectancy (from birth): 74.5

Total Fertility Rate (average children per woman): 2.7

Religion: nominally Roman Catholic 90% (less than 20% practicing), Protestant 2%, Jewish 2%, other 6%

Languages: Spanish (official), English, Italian, German, French

Literacy: 96.2%

GDP Per Capita: $8,600

Occupations: services 57%, industry 31%, agriculture 12% (1985 est.)

Climate: mostly temperate; arid in southeast; sub-antarctic in southwest

Terrain: rich plains of the Pampas in northern half; flat to rolling plateau of Patagonia in south; rugged Andes along western border

Armenia

Long Name: Republic of Armenia

Location: Southwestern Asia, east of Turkey

Capital: Yerevan

Government Type: republic

Currency: dram

Area: 10,962 sq. mi. (28,400 sq. km)

Population: 3,409,000

Population Growth Rate: -0.4%

Birth Rate (per 1,000): 14

Death Rate (per 1,000): 9

Life Expectancy (from birth): 66.7

Total Fertility Rate (average children per woman): 1.7

Religion: Armenian Orthodox 94%

Languages: Armenian 96%, Russian 2%, other 2%

Literacy: 99%

GDP Per Capita: $2,800

Occupations: agriculture 38%, services 37%, industry and construction 23%, other 2%

Climate: highland continental; hot summers, cold winters

Terrain: high Armenian Plateau with mountains; little forest land; fast-flowing rivers; good soil in Aras River valley

Australia

Long Name: Commonwealth of Australia

Location: Oceania, continent between the Indian Ocean and the South Pacific Ocean

Capital: Canberra

Government Type: federal parliamentary state

Currency: Australian dollar ($A)

Area: 2,940,521 sq. mi. (7,617,930 sq. km)

Population: 18,784,000

Population Growth Rate: 0.9%

Birth Rate (per 1,000): 13

Death Rate (per 1,000): 7

Life Expectancy (from birth): 79.9

Total Fertility Rate (average children per woman): 1.8

Religion: Anglican 26.1%, Roman Catholic 26%, other Christian 24.3%

Languages: English, native languages

Literacy: 100%

GDP Per Capita: $23,600

Occupations: finance and services 34%, public and community services 23%, wholesale and retail trade 20%, manufacturing and industry 17%, agriculture 6% (1987 est.)

Climate: generally arid to semiarid; temperate in south and east; tropical in north

Terrain: mostly low plateau with deserts; fertile plain in southeast

Austria

Long Name: Republic of Austria

Location: Central Europe, north of Italy and Slovenia

Capital: Vienna

Government Type: federal republic

Currency: Austrian schilling (AS)

Area: 31,934 sq. mi. (82,730 sq. km)

Population: 8,139,000

Population Growth Rate: 0.1%

Birth Rate (per 1,000): 10

Death Rate (per 1,000): 10

Life Expectancy (from birth): 77.3

311

Total Fertility Rate (average children per woman): 1.4

Religion: Roman Catholic 85%, Protestant 6%, other 9%

Languages: German

Literacy: 99%

GDP Per Capita: $19,700

Occupations: services 56.4%, industry and crafts 35.4%, agriculture and forestry 8.1%

Climate: temperate; continental, cloudy; cold winters with frequent rain in lowlands and snow in mountains; cool summers with occasional showers

Terrain: in the west and south mostly mountains (Alps); along the eastern and northern margins mostly flat or gently sloping

Azerbaijan

Long Name: Azerbaijani Republic

Location: Southwestern Asia, bordering the Caspian Sea, between Iran and Russia

Capital: Baku (Baki)

Government Type: republic

Currency: manat

Area: 33,235 sq. mi. (86,100 sq. km)

Population: 7,908,000

Population Growth Rate: 0.7%

Birth Rate (per 1,000): 22

Death Rate (per 1,000): 9

Life Expectancy (from birth): 63.3

Total Fertility Rate (average children per woman): 2.7

Religion: Muslim 93.4%, Russian Orthodox 2.5%, Armenian Orthodox 2.3%, other 1.8%

Languages: Azeri 89%, Russian 3%, Armenian 2%, other 6%

Literacy: 99%

GDP Per Capita: $1,550

Occupations: agriculture and forestry 32%, industry and construction 26%, other 42%

Climate: dry, semiarid steppe

Terrain: large, flat Kur-Araz Lowland (much of it below sea level) with Great Caucasus Mountains to the north; Qarabag (Karabakh) Upland in west; Baku lies on Abseron (Apsheron) Peninsula that juts into Caspian Sea

The Bahamas

Long Name: Commonwealth of The Bahamas

Location: Caribbean, chain of islands in the North Atlantic Ocean, southeast of Florida

Capital: Nassau

Government Type: commonwealth

Currency: Bahamian dollar (B$)

Area: 3,887 sq. mi. (10,070 sq. km)

Population: 283,705

Population Growth Rate: 1.4%

Birth Rate (per 1,000): 21

Death Rate (per 1,000): 5

Life Expectancy (from birth): 74.0

Total Fertility Rate (average children per woman): 2.3

Religion: Baptist 32%, Anglican 20%, Roman Catholic 19%, Methodist 6%, Church of God 6%, other Protestant 12%, none or unknown 3%, other 2%

Languages: English, Creole

Literacy: 98.2%

GDP Per Capita: $18,700

Occupations: tourism 40%, government 30%, business services 10%, agriculture 5%

Climate: tropical marine; moderated by warm waters of Gulf Stream

Terrain: long, flat coral formations with some low, rounded hills

Bahrain

Long Name: State of Bahrain

Location: Middle East, archipelago in the Persian Gulf, east of Saudi Arabia

Capital: Manama

Government Type: traditional monarchy

Currency: Bahraini dinar (BD)

Area: 239 sq. mi. (620 sq. km)

Population: 629,090

Population Growth Rate: 2.1%

Birth Rate (per 1,000): 22

Death Rate (per 1,000): 3

Life Expectancy (from birth): 75.0

Total Fertility Rate (average children per woman): 3.0

Religion: Shi'a Muslim 75%, Sunni Muslim 25%

Languages: Arabic, English, Farsi, Urdu

Literacy: 85.2%

GDP Per Capita: $13,000

Occupations: industry and commerce 85%, agriculture 5%, services 5%, government 3% (1982 est.)

Climate: arid; mild, pleasant winters; very hot, humid summers

Terrain: mostly low desert plain rising gently to low central escarpment

Bangladesh

Long Name: People's Republic of Bangladesh

Location: Southern Asia, bordering the Bay of Bengal, between Burma and India

Capital: Dhaka

Government Type: republic

Currency: taka (Tk)

Area: 51,689 sq. mi. (133,910 sq. km)

Population: 129,798,000

Population Growth Rate: 1.8%

Birth Rate (per 1,000): 29

Death Rate (per 1,000): 11

Life Expectancy (from birth): 56.7

Total Fertility Rate (average children per woman): 3.3

Religion: Muslim 88.3%, Hindu 10.5%, other 1.2%

Languages: Bangla (official), English

Literacy: 38.1%

GDP Per Capita: $1,260

Occupations: agriculture 65%, services 21%, industry and mining 14% (1989 est.)

Climate: tropical; cool, dry winter (October to March); hot, humid summer (March to June); cool, rainy monsoon (June to October)

Terrain: mostly flat alluvial plain; hilly in southeast

Barbados

Long Name: Barbados

Location: Caribbean, island between the Caribbean Sea and the North Atlantic Ocean, northeast of Venezuela

Capital: Bridgetown

Government Type: parliamentary democracy

Currency: Barbadian dollar (Bds$)

Area: 166 sq. mi. (430 sq. km)

Population: 259,191

Population Growth Rate: 0.1%

Birth Rate (per 1,000): 15

Death Rate (per 1,000): 8

Life Expectancy (from birth): 74.8

Total Fertility Rate (average children per woman): 1.9

Religion: Protestant 67% (Anglican 40%, Pentecostal 8%, Methodist 7%, other 12%), Roman Catholic 4%, none 17%, unknown 3%, other 9% (1980 est.)

Languages: English

Literacy: 97.4%

GDP Per Capita: $10,300

Occupations: services and government 41%, manufacturing and construction 18%, commerce 15%, transportation, storage, communications, and financial institutions 8%, agriculture 6%, utilities 2%

Climate: tropical; rainy season (June to October)

Terrain: relatively flat; rises gently to central highland region

Belarus

Long Name: Republic of Belarus

Location: Eastern Europe, east of Poland

Capital: Minsk

Government Type: republic

Currency: Belarusian ruble (BR)

Area: 80,134 sq. mi. (207,600 sq. km)

Population: 10,402,000

Population Growth Rate: -0.1%

Birth Rate (per 1,000): 10

Death Rate (per 1,000): 13

Life Expectancy (from birth): 68.3

Total Fertility Rate (average children per woman): 1.3

Religion: Eastern Orthodox 80%, other (including Roman Catholic, Protestant, Jewish, and Muslim) 20%

Languages: Byelorussian and Russian

Literacy: 98%

GDP Per Capita: $5,000

Occupations: industry and construction 36%, agriculture and forestry 19%, services 45%

Climate: cold winters, cool and moist summers; transitional between continental and maritime regions

Terrain: generally flat and contains much marshland

Belgium

Long Name: Kingdom of Belgium

Location: Western Europe, bordering the North Sea, between France and the Netherlands

Capital: Brussels

Government Type: federal parliamentary democracy under a constitutional monarch

Currency: Belgian franc (BF)

Area: 11,669 sq. mi. (30,230 sq. km)

Population: 10,182,000

Population Growth Rate: 0.1%

Birth Rate (per 1,000): 10

Death Rate (per 1,000): 10

Life Expectancy (from birth): 77.4

Total Fertility Rate (average children per woman): 1.5

Religion: Roman Catholic 75%, Protestant or other 25%

Languages: Flemish 56%, French 32%, German 1%, legally bilingual 11%

Literacy: 99%

GDP Per Capita: $20,300

Occupations: services 69.7%, industry 27.7%, agriculture 2.6%

Climate: temperate; mild winters, cool summers; rainy, humid, cloudy

Terrain: flat coastal plains in northwest; central rolling hills; rugged mountains of Ardennes Forest in southeast

Belize

Long Name: Belize

Location: Central America, bordering the Caribbean Sea, between Guatemala and Mexico

Capital: Belmopan

Government Type: parliamentary democracy

Currency: Belizean dollar (Bz$)

Area: 8,801 sq. mi. (22,800 sq. km)

Population: 235,789

Population Growth Rate: 2.4%

Birth Rate (per 1,000): 31

Death Rate (per 1,000): 6

Life Expectancy (from birth): 69.0

Total Fertility Rate (average children per woman): 3.9

Religion: Roman Catholic 62%, Protestant 30% (Anglican 12%, Methodist 6%, Mennonite 4%, Seventh-Day Adventist 3%, Pentecostal 2%, Jehovah's Witnesses 1%, other 2%), none 2%, other 6%

Languages: English (official), Spanish, Mayan, Garifuna (Carib)

Literacy: 70.3%

GDP Per Capita: $2,960

Occupations: agriculture 30%, services 16%, government 15.4%, commerce 11.2%, manufacturing 10.3%

Climate: tropical; very hot and humid; rainy season (May to February)

Terrain: flat, swampy coastal plain; low mountains in south

Benin

Long Name: Republic of Benin

Location: Western Africa, bordering the North Atlantic Ocean, between Nigeria and Togo

Capital: Porto-Novo

Government Type: republic under multiparty democratic rule

Currency: Communaute Financiere Africaine franc (CFAF)

Area: 42,699 sq. mi. (110,620 sq. km)

Population: 6,306,000

Population Growth Rate: 3.3%

Birth Rate (per 1,000): 46

Death Rate (per 1,000): 13

Life Expectancy (from birth): 53.6

Total Fertility Rate (average children per woman): 6.5

Religion: indigenous beliefs 70%, Muslim 15%, Christian 15%

Languages: French (official), Fon and Yoruba, tribal languages

Literacy: 37%

GDP Per Capita: $1,440

Occupations: agriculture

Climate: tropical; hot, humid in south; semiarid in north

Terrain: mostly flat to undulating plain; some hills and low mountains

Bhutan

Long Name: Kingdom of Bhutan

Location: Southern Asia, between China and India

Capital: Thimphu

Government Type: monarchy

Currency: ngultrum (Nu)

Area: 18,142 sq. mi. (47,000 sq. km)

Population: 1,952,000

Population Growth Rate: 2.3%

Birth Rate (per 1,000): 37

Death Rate (per 1,000): 15

Life Expectancy (from birth): 52.3

Total Fertility Rate (average children per woman): 5.2

Religion: Lamaistic Buddhism 75%, Indian- and Nepalese-influenced Hinduism 25%

Languages: Dzongkha (official)

Literacy: 42.2%

GDP Per Capita: $730

Occupations: agriculture 93%, services 5%, industry and commerce 2%

Climate: varies; tropical in southern plains; cool winters and hot summers in central valleys; severe winters and cool summers in Himalayas

Terrain: mostly mountainous with some fertile valleys and savanna

Bolivia

Long Name: Republic of Bolivia

Location: Central South America, southwest of Brazil

Capital: La Paz (seat of government); Sucre (legal capital and seat of judiciary)

Government Type: republic

Currency: boliviano ($B)

Area: 418,575 sq. mi. (1,084,390 sq. km)

Population: 7,983,000

Population Growth Rate: 2%

Birth Rate (per 1,000): 31

Death Rate (per 1,000): 10

Life Expectancy (from birth): 60.9

Total Fertility Rate (average children per woman): 4.1

Religion: Roman Catholic 95%, Protestant (Evangelical Methodist)

Languages: Spanish (official), Quechua (official), Aymara (official)

Literacy: 83.1%

GDP Per Capita: $3,000

Occupations: agriculture, services and utilities, manufacturing, mining and construction

Climate: varies with altitude; humid and tropical to cold and semiarid

Terrain: rugged Andes Mountains with a highland plateau (Altiplano); hills; lowland plains of the Amazon Basin

Bosnia and Herzegovina

Long Name: Bosnia and Herzegovina

Location: Southeastern Europe, bordering the Adriatic Sea and Croatia

Capital: Sarajevo

Government Type: emerging democracy

Currency: dinar

Area: 19,776 sq. mi. (51,233 sq. km)

Population: 3,482,000

Population Growth Rate: 3.6%

Birth Rate (per 1,000): 9

Death Rate (per 1,000): 12

Life Expectancy (from birth): 63.0

Total Fertility Rate (average children per woman): 1.1

Religion: Muslim 40%, Orthodox 31%, Catholic 15%, Protestant 4%, other 10%

Languages: Serbo-Croatian (often called Bosnian) 99%

Literacy: 90%

GDP Per Capita: $600

Occupations: industry, mining

Climate: hot summers and cold winters; areas of high elevation have short, cool summers and long, severe winters; mild, rainy winters along coast

Terrain: mountains and valleys

Botswana

Long Name: Republic of Botswana

Location: Southern Africa, north of South Africa

Capital: Gaborone

Government Type: parliamentary republic

Currency: pula

Area: 225,953 sq. mi. (585,370 sq. km)

Population: 1,464,000

Population Growth Rate: 1.1%

Birth Rate (per 1,000): 32

Death Rate (per 1,000): 21

Life Expectancy (from birth): 40.1

Total Fertility Rate (average children per woman): 4.0

Religion: indigenous beliefs 50%, Christian 50%

Languages: English (official), Setswana

Literacy: 69.8%

GDP Per Capita: $3,100

Occupations: mining, agriculture

Climate: semiarid; warm winters and hot summers

Terrain: predominately flat to gently rolling tableland; Kalahari Desert in southwest

Brazil

Long Name: Federative Republic of Brazil

Location: Eastern South America, bordering the Atlantic Ocean

Capital: Brasilia

Government Type: federal republic

Currency: real (R$)

Area: 3,264,213 sq. mi. (8,456,510 sq. km)

Population: 171,853,000

Population Growth Rate: 1.2%

Birth Rate (per 1,000): 21

Death Rate (per 1,000): 9

Life Expectancy (from birth): 64.4

Total Fertility Rate (average children per woman): 2.3

Religion: Roman Catholic (nominal) 70%

Languages: Portuguese (official), Spanish, English, French

Literacy: 83.3%

GDP Per Capita: $6,300

Occupations: services 42%, agriculture 31%, industry 27%

Climate: mostly tropical; temperate in south

Terrain: mostly flat to rolling lowlands in north; some plains, hills, mountains, and narrow coastal belt

Brunei

Long Name: Negara Brunei Darussalam

Location: Southeastern Asia, bordering the South China Sea and Malaysia

Capital: Bandar Seri Begawan

Government Type: constitutional sultanate

Currency: Bruneian dollar (B$)

Area: 2,034 sq. mi. (5,270 sq. km)

Population: 322,982

Population Growth Rate: 2.4%

Birth Rate (per 1,000): 25

Death Rate (per 1,000): 5

Life Expectancy (from birth): 71.7

Total Fertility Rate (average children per woman): 3.4

Religion: Muslim (official) 63%, Buddhism 14%, Christian 8%, indigenous beliefs and other 15% (1981 est.)

Languages: Malay (official), English, Chinese

321

Literacy: 88.2%

GDP Per Capita: $15,800

Occupations: government 48%, production of oil, natural gas, services, and construction 42%, agriculture, forestry, and fishing 4%, other 6% (1986 est.)

Climate: tropical; hot, humid, rainy

Terrain: flat coastal plain rises to mountains in east; hilly lowland in west

Bulgaria

Long Name: Republic of Bulgaria

Location: Southeastern Europe, bordering the Black Sea, between Romania and Turkey

Capital: Sofia

Government Type: emerging democracy

Currency: lev (Lv)

Area: 42,672 sq. mi. (110,550 sq. km)

Population: 8,195,000

Population Growth Rate: -0.6%

Birth Rate (per 1,000): 8

Death Rate (per 1,000): 13

Life Expectancy (from birth): 72.0

Total Fertility Rate (average children per woman): 1.1

Religion: Bulgarian Orthodox 85%, Muslim 13%, Jewish 0.8%, Roman Catholic 0.5%, Uniate Catholic 0.2%, Protestant, Gregorian-Armenian, and other 0.5%

Languages: Bulgarian

Literacy: 98%

GDP Per Capita: $4,630

Occupations: industry 41%, agriculture 18%, other 41%

Climate: temperate; cold, damp winters; hot, dry summers

Terrain: mostly mountains with lowlands in north and southeast

Burkina Faso

Long Name: Burkina Faso

Location: Western Africa, north of Ghana

Capital: Ouagadougou

Government Type: parliamentary

Currency: Communaute Financiere Africaine franc (CFAF)

Area: 105,687 sq. mi. (273,800 sq. km)

Population: 11,576,000

Population Growth Rate: 2.7%

Birth Rate (per 1,000): 46

Death Rate (per 1,000): 18

Life Expectancy (from birth): 46.1

Total Fertility Rate (average children per woman): 6.6

Religion: Muslim 50%, indigenous beliefs 40%, Christian (mainly Roman Catholic) 10%

Languages: French (official), tribal languages belonging to Sudanic family, spoken by 90% of the population

Literacy: 19.2%

GDP Per Capita: $740

Occupations: agriculture 80%, industry 15%, commerce, services, and government 5%

Climate: tropical; warm, dry winters; hot, wet summers

Terrain: mostly flat to dissected, undulating plains; hills in west and southeast

Burundi

Long Name: Republic of Burundi

Location: Central Africa, east of Democratic Republic of the Congo

Capital: Bujumbura

Government Type: republic

Currency: Burundi franc (FBu)

Area: 9,901 sq. mi. (25,650 sq. km)

Population: 5,736,000

Population Growth Rate: 3.5%

Birth Rate (per 1,000): 42

Death Rate (per 1,000): 17

Life Expectancy (from birth): 45.6

Total Fertility Rate (average children per woman): 6.4

Religion: Christian 67% (Roman Catholic 62%, Protestant 5%), indigenous beliefs 32%, Muslim 1%

Languages: Kirundi (official), French (official), Swahili (along Lake Tanganyika and in the Bujumbura area)

Literacy: 35.3%

GDP Per Capita: $600

Occupations: agriculture 93%, government 4%, industry and commerce 1.5%, services 1.5% (1983 est.)

Climate: equatorial; high plateau with considerable altitude variation; average annual temperature varies with altitude; wet seasons from February to May and September to November, and dry seasons from June to August and December to January

Terrain: hilly and mountainous, dropping to a plateau in east; some plains

Cambodia

Long Name: Kingdom of Cambodia

Location: Southeastern Asia, bordering the Gulf of Thailand, between Thailand and Vietnam

Capital: Phnom Penh

Government Type: multiparty democracy under constitutional monarchy

Currency: new riel (CR)

Area: 68,137 sq. mi. (176,520 sq. km)

Population: 11,627,000

Population Growth Rate: 2.5%

Birth Rate (per 1,000): 42

Death Rate (per 1,000): 16

Life Expectancy (from birth): 48.0

Total Fertility Rate (average children per woman): 5.8

Religion: Theravada Buddhism 95%, other 5%

Languages: Khmer (official), French

Literacy: 35%

GDP Per Capita: $710

Occupations: agriculture 80%

Climate: tropical; rainy, monsoon season (May to November); dry season (December to April); little seasonal temperature variation

Terrain: mostly low, flat plains; mountains in southwest and north

Cameroon

Long Name: Republic of Cameroon

Location: Western Africa, bordering the Bight of Biafra, between Equatorial Guinea and Nigeria

Capital: Yaounde

Government Type: unitary republic; multiparty presidential regime

Currency: Communaute Financiere Africaine franc (CFAF)

Area: 181,204 sq. mi. (469,440 sq. km)

Population: 15,456,000

Population Growth Rate: 2.8%

Birth Rate (per 1,000): 42

Death Rate (per 1,000): 14

Life Expectancy (from birth): 51.4

Total Fertility Rate (average children per woman): 5.9

Religion: indigenous beliefs 51%, Christian 33%, Muslim 16%

Languages: 24 major African language groups, English (official), French (official)

Literacy: 63.4%

GDP Per Capita: $1,230

Occupations: oil, agriculture

Climate: varies with climate, from tropical along coast to semiarid and hot in north

Terrain: diverse, with coastal plain in southwest, dissected plateau in center, mountains in west, plains in north

Canada

Long Name: Canada

Location: Northern North America, bordering the North Atlantic Ocean and North Pacific Ocean, north of the conterminous US

Capital: Ottawa

Government Type: confederation with parliamentary democracy

Currency: Canadian dollar (Can$)

Area: 3,559,294 sq. mi. (9,220,970 sq. km)

Population: 31,006,000

Population Growth Rate: 1.1%

Birth Rate (per 1,000): 12

Death Rate (per 1,000): 7

Life Expectancy (from birth): 79.2

Total Fertility Rate (average children per woman): 1.7

Religion: Roman Catholic 45%, United Church 12%, Anglican 8%, other 35%

Languages: English (official), French (official)

Literacy: 97%

GDP Per Capita: $25,000

Occupations: services 74%, manufacturing 15%, agriculture 3%, construction 5%, other 3%

Climate: varies from temperate in south to subarctic and arctic in north

Terrain: mostly plains, with mountains in west and lowlands in southeast

Cape Verde

Long Name: Republic of Cape Verde

Location: Western Africa, group of islands in the North Atlantic Ocean, west of Senegal

Capital: Praia

Government Type: republic

Currency: Cape Verdean escudo (CVEsc)

Area: 1,556 sq. mi. (4,030 sq. km)

Population: 405,748

Population Growth Rate: 1.5%

Birth Rate (per 1,000): 34

Death Rate (per 1,000): 7

Life Expectancy (from birth): 70.5

Total Fertility Rate (average children per woman): 5.1

Religion: Roman Catholicism fused with indigenous beliefs

Languages: Portuguese, Crioulo (a blend of Portuguese and West African words)

Literacy: 71.6%

GDP Per Capita: $1,000

Occupations: agriculture

Climate: temperate; warm, dry summer; precipitation meager and very erratic

Terrain: steep, rugged, rocky, volcanic

Central African Republic

Long Name: Central African Republic

Location: Central Africa, north of Democratic Republic of the Congo

Capital: Bangui

Government Type: republic

Currency: Communaute Financiere Africaine franc (CFAF)

Area: 240,470 sq. mi. (622,980 sq. km)

Population: 3,445,000

Population Growth Rate: 2%

Birth Rate (per 1,000): 39

Death Rate (per 1,000): 17

Life Expectancy (from birth): 46.8

Total Fertility Rate (average children per woman): 5.1

Religion: Protestant 25%, Roman Catholic 25%, indigenous beliefs 24%, Muslim 15%, other 11%

Languages: French (official), Sangho (lingua franca and national language), Arabic, Hunsa, Swahili

Literacy: 60%

GDP Per Capita: $800

Occupations: agriculture, industry, mining

Climate: tropical; hot, dry winters; mild to hot, wet summers

Terrain: vast, flat to rolling, monotonous plateau; scattered hills in northeast and south-west

Chad

Long Name: Republic of Chad

Location: Central Africa, south of Libya

Capital: N'Djamena

Government Type: republic

Currency: Communaute Financiere Africaine franc (CFAF)

Area: 486,051 sq. mi. (1,259,200 sq. km)

Population: 7,557,000

Population Growth Rate: 2.7%

Birth Rate (per 1,000): 43

Death Rate (per 1,000): 17

Life Expectancy (from birth): 48.2

Total Fertility Rate (average children per woman): 5.7

Religion: Muslim 50%, Christian 25%, indigenous beliefs (mostly animism) 25%

Languages: French (official), Arabic (official), Sara and Sango (in south), more than 100 different languages and dialects

Literacy: 48.1%

GDP Per Capita: $600

Occupations: agriculture 85%

Climate: tropical in south, desert in north

Terrain: broad, arid plains in center; desert in north; mountains in northwest; lowlands in south

Chile

Long Name: Republic of Chile

Location: Southern South America, bordering the South Atlantic Ocean and South Pacific Ocean, between Argentina and Peru

Capital: Santiago

Government Type: republic

Currency: Chilean peso (Ch$)

Area: 289,037 sq. mi. (748,800 sq. km)

Population: 14,974,000

Population Growth Rate: 1.3%

Birth Rate (per 1,000): 18

Death Rate (per 1,000): 6

Life Expectancy (from birth): 75.2

Total Fertility Rate (average children per woman): 2.3

Religion: Roman Catholic 89%, Protestant 11%, Jewish less than 1%

Languages: Spanish

Literacy: 95.2%

GDP Per Capita: $8,400

Occupations: services 38.3% (includes government 12%), industry and commerce 33.8%, agriculture, forestry, and fishing 19.2%, construction 6.4%, mining 2.3%

Climate: temperate; desert in north; cool and damp in south

Terrain: low coastal mountains; fertile central valley; rugged Andes in east

China

Long Name: People's Republic of China

Location: Eastern Asia, bordering the East China Sea, Korea Bay, Yellow Sea, and South China Sea, between North Korea and Vietnam

Capital: Beijing

Government Type: Communist state

Currency: yuan (´)

Area: 3,599,994 sq. mi. (9,326,410 sq. km)

Population: 1,246,872,000

Population Growth Rate: 0.8%

Birth Rate (per 1,000): 16

Death Rate (per 1,000): 7

Life Expectancy (from birth): 69.6

Total Fertility Rate (average children per woman): 1.8

Religion: Officially atheist but Daoism (Taoism), Buddhism, Muslim, and Christian

Languages: Standard Chinese or Mandarin (Putonghua, based on the Beijing dialect), Yue (Cantonese), Wu (Shanghaiese), Minbei (Fuzhou), Minnan (Hokkien-Taiwanese), Xiang, Gan, Hakka dialects

Literacy: 81.5%

GDP Per Capita: $2,800

Occupations: agriculture and forestry 54%, industry and commerce 26%, construction and mining 7%, social services 6%, other 7%

Climate: extremely diverse; tropical in south to subarctic in north

Terrain: mostly mountains, high plateaus, deserts in west; plains, deltas, and hills in east

Colombia

Long Name: Republic of Colombia

Location: Northern South America, bordering the Caribbean Sea, between Panama and Venezuela, and bordering the North Pacific Ocean, between Ecuador and Panama

Capital: Bogota

Government Type: republic; executive branch dominates government structure

Currency: Colombian peso (Col$)

Area: 400,938 sq. mi. (1,038,700 sq. km)

Population: 39,309,000

Population Growth Rate: 1.9%

Birth Rate (per 1,000): 25

Death Rate (per 1,000): 6

Life Expectancy (from birth): 70.1

Total Fertility Rate (average children per woman): 2.9

Religion: Roman Catholic 95%

Languages: Spanish

Literacy: 91.3%

GDP Per Capita: $5,400

Occupations: services 46%, agriculture 30%, industry 24%

Climate: tropical along coast and eastern plains; cooler in highlands

Terrain: flat coastal lowlands; central highlands; high Andes Mountains; eastern lowland plains

Comoros

Long Name: Federal Islamic Republic of the Comoros

Location: Southern Africa, group of islands in the Mozambique Channel, about two-thirds of the way between northern Madagascar and northern Mozambique

Capital: Moroni

Government Type: independent republic

Currency: Comoran franc (CF)

Area: 838 sq. mi. (2,170 sq. km)

Population: 562,723

Population Growth Rate: 3.1%

Birth Rate (per 1,000): 41

Death Rate (per 1,000): 10

Life Expectancy (from birth): 60.4

Total Fertility Rate (average children per woman): 5.5

Religion: Sunni Muslim 86%, Roman Catholic 14%

Languages: Arabic (official), French (official), Comoran (a blend of Swahili and Arabic)

Literacy: 57.3%

GDP Per Capita: $650

Occupations: agriculture 80%, government 3%

Climate: tropical marine; rainy season (November to May)

Terrain: volcanic islands, interiors vary from steep mountains to low hills

Democratic Republic of the Congo

Long Name: Democratic Republic of the Congo

Location: Central Africa, northeast of Angola

Capital: Kinshasa

330   **Government Type:** republic with a strong presidential system

Currency: Congolese franc

Area: 875,294 sq. mi. (2,267,600 sq. km)

Population: 50,481,000

Population Growth Rate: 3%

Birth Rate (per 1,000): 47

Death Rate (per 1,000): 15

Life Expectancy (from birth): 49.3

Total Fertility Rate (average children per woman): 6.5

Religion: Roman Catholic 50%, Protestant 20%, Kimbanguist 10%, Muslim 10%, other syncretic sects and traditional beliefs 10%

Languages: French (official), Lingala (a lingua franca trade language), Kingwana (a dialect of Kiswahili or Swahili), Kikongo, Tshiluba

Literacy: 77.3%

GDP Per Capita: $400

Occupations: agriculture 65%, industry 16%, services 19%

Climate: tropical; hot and humid in equatorial river basin; cooler and drier in southern highlands; cooler and wetter in eastern highlands; north of Equator—wet season April to October, dry season December to February; south of Equator—wet season November to March, dry season April to October

Terrain: vast central basin is a low-lying plateau; mountains in east

Republic of the Congo

Long Name: Republic of the Congo

Location: Western Africa, bordering the South Atlantic Ocean, between the Democratic Republic of the Congo and Gabon

Capital: Brazzaville

Government Type: republic

Currency: Communaute Financiere Africaine franc (CFAF)

Area: 131,819 sq. mi. (341,500 sq. km)

Population: 2,717,000

Population Growth Rate: 2.2%

Birth Rate (per 1,000): 39

Death Rate (per 1,000): 16

Life Expectancy (from birth): 47.1

Total Fertility Rate (average children per woman): 5.0

Religion: Christian 50%, animist 48%, Muslim 2%

331

Languages: French (official), African languages (Lingala and Kikongo are the most widely used)

Literacy: 74.9%

GDP Per Capita: $1,960

Occupations: agriculture, oil

Climate: tropical; rainy season (March to June); dry season (June to October); constantly high temperatures and humidity

Terrain: coastal plain; southern basin; central plateau; northern basin

Costa Rica

Long Name: Republic of Costa Rica

Location: Central America, bordering both the Caribbean Sea and the North Pacific Ocean, between Nicaragua and Panama

Capital: San Jose

Government Type: democratic republic

Currency: Costa Rican colon (C)

Area: 19,554 sq. mi. (50,660 sq. km)

Population: 3,674,000

Population Growth Rate: 1.9%

Birth Rate (per 1,000): 23

Death Rate (per 1,000): 4

Life Expectancy (from birth): 75.9

Total Fertility Rate (average children per woman): 2.8

Religion: Roman Catholic 95%

Languages: Spanish

Literacy: 94.8%

GDP Per Capita: $5,500

Occupations: industry and commerce 35.1%, government and services 33%, agriculture 27%, other 4.9% (1985 est.)

Climate: tropical; dry season (December to April); rainy season (May to November)

Terrain: coastal plains separated by rugged mountains

Cote d'Ivoire

Long Name: Republic of Cote d'Ivoire

Location: Western Africa, bordering the North Atlantic Ocean, between Ghana and Liberia

Capital: Yamoussoukro (official) and Abidjan (de facto)

Government Type: republic

Currency: Communaute Financiere Africaine franc (CFAF)

Area: 122,748 sq. mi. (318,000 sq. km)

Population: 15,818,000

Population Growth Rate: 2.4%

Birth Rate (per 1,000): 42

Death Rate (per 1,000): 16

Life Expectancy (from birth): 46.2

Total Fertility Rate (average children per woman): 6.0

Religion: Muslim 60%, indigenous 25%, Christian 12%

Languages: French (official), 60 native dialects with Dioula the most widely spoken

Literacy: 40.1%

GDP Per Capita: $1,620

Occupations: agriculture and forestry

Climate: tropical along coast, semiarid in far north; three seasons—warm and dry (November to March), hot and dry (March to May), hot and wet (June to October)

Terrain: mostly flat to undulating plains; mountains in northwest

Croatia

Long Name: Republic of Croatia

Location: Southeastern Europe, bordering the Adriatic Sea, between Bosnia and Herzegovina and Slovenia

Capital: Zagreb

Government Type: presidential/parliamentary democracy

Currency: Croatian kuna (HRK)

Area: 21,774 sq. mi. (56,410 sq. km)

Population: 4,677,000

Population Growth Rate: 0.1%

Birth Rate (per 1,000): 10

Death Rate (per 1,000): 11

Life Expectancy (from birth): 73.8

Total Fertility Rate (average children per woman): 1.5

Religion: Catholic 76.5%, Orthodox 11.1%, Slavic Muslim 1.2%, Protestant 0.4%, others and unknown 10.8%

333

Languages: Serbo-Croatian 96%, other 4% (including Italian, Hungarian, Czechoslovak, and German)

Literacy: 97%

GDP Per Capita: $4,300

Occupations: industry and mining 31.1%, government 19.1% (including education and health), agriculture 4.3%, other 45.5%

Climate: Mediterranean and continental; continental climate predominant with hot summers and cold winters; mild winters, dry summers along coast

Terrain: geographically diverse; flat plains along Hungarian border, low mountains and highlands near Adriatic coast, coastline, and islands

Cuba

Long Name: Republic of Cuba

Location: Caribbean, island between the Caribbean Sea and the North Atlantic Ocean, south of Florida

Capital: Havana

Government Type: Communist state

Currency: Cuban peso (Cu$)

Area: 42,792 sq. mi. (110,860 sq. km)

Population: 11,096,000

Population Growth Rate: 0.4%

Birth Rate (per 1,000): 13

Death Rate (per 1,000): 7

Life Expectancy (from birth): 75.6

Total Fertility Rate (average children per woman): 1.6

Religion: nominally Roman Catholic prior to Communism

Languages: Spanish

Literacy: 95.7%

GDP Per Capita: $1,480

Occupations: services and government 30%, industry 22%, agriculture 20%, commerce 11%, construction 10%, transportation and communications 7%

Climate: tropical; moderated by trade winds; dry season (November to April); rainy season (May to October)

Terrain: mostly flat to rolling plains with rugged hills and mountains in the southeast

Cyprus

Long Name: Republic of Cyprus

Location: Middle East, island in the Mediterranean Sea, south of Turkey

Capital: Nicosia

Government Type: republic

Currency: Cypriot pound (£C)

Area: 3,567 sq. mi. (9,240 sq. km)

Population: 754,064

Population Growth Rate: 0.7%

Birth Rate (per 1,000): 14

Death Rate (per 1,000): 8

Life Expectancy (from birth): 76.8

Total Fertility Rate (average children per woman): 2.0

Religion: Greek Orthodox 78%, Muslim 18%, Maronite, Armenian Apostolic, and other 4%

Languages: Greek, Turkish, English

Literacy: 94%

GDP Per Capita: $11,800

Occupations: services 62%, industry 25%, agriculture 13%

Climate: temperate; Mediterranean with hot, dry summers and cool, wet winters

Terrain: central plain with mountains to north and south; scattered but significant plains along southern coast

Czech Republic

Long Name: Czech Republic

Location: Central Europe, southeast of Germany

Capital: Prague

Government Type: parliamentary democracy

Currency: koruna (Kc)

Area: 30,357 sq. mi. (78,645 sq. km)

Population: 10,281,000

Population Growth Rate: -0.1%

Birth Rate (per 1,000): 9

Death Rate (per 1,000): 11

335

Life Expectancy (from birth): 74.1

Total Fertility Rate (average children per woman): 1.2

Religion: atheist 39.8%, Roman Catholic 39.2%, Protestant 4.6%, Orthodox 3%, other 13.4%

Languages: Czech, Slovak

Literacy: 99%

GDP Per Capita: $11,100

Occupations: services 43.7%, industry 33.1%, agriculture 6.9%, construction 9.1%, transport and communications 7.2%

Climate: temperate; cool summers; cold, cloudy, humid winters

Terrain: Bohemia in the west consists of rolling plains, hills, and plateaus surrounded by low mountains; Moravia in the east consists of very hilly country

Denmark

Long Name: Kingdom of Denmark

Location: Northern Europe, bordering the Baltic Sea and the North Sea, on a peninsula north of Germany

Capital: Copenhagen

Government Type: constitutional monarchy

Currency: Danish krone (DKr)

Area: 16,364 sq. mi. (42,394 sq. km)

Population: 5,357,000

Population Growth Rate: 0.5%

Birth Rate (per 1,000): 12

Death Rate (per 1,000): 11

Life Expectancy (from birth): 76.3

Total Fertility Rate (average children per woman): 1.7

Religion: Evangelical Lutheran 91%, other Protestant and Roman Catholic 2%, other 7%

Languages: Danish, Faroese, Greenlandic (an Eskimo dialect)

Literacy: 99%

GDP Per Capita: $22,700

Occupations: private services 40%, government services 30%, manufacturing and mining 19%, construction 6%, agriculture, forestry, and fishing 5%

Climate: temperate; humid and overcast; mild, windy winters and cool summers

Terrain: low and flat to gently rolling plains

Djibouti

Long Name: Republic of Djibouti

Location: Eastern Africa, bordering the Gulf of Aden and the Red Sea, between Eritrea and Somalia

Capital: Djibouti

Government Type: republic

Currency: Djiboutian franc (DF)

Area: 8,484 sq. mi. (21,980 sq. km)

Population: 447,439

Population Growth Rate: 1.5%

Birth Rate (per 1,000): 42

Death Rate (per 1,000): 15

Life Expectancy (from birth): 51.1

Total Fertility Rate (average children per woman): 5.9

Religion: Muslim 94%, Christian 6%

Languages: French (official), Arabic (official), Somali, Afar

Literacy: 46.2%

GDP Per Capita: $1,200

Occupations: agriculture 75%, services 14%, industry 11%

Climate: desert; torrid, dry

Terrain: coastal plain and plateau separated by central mountains

Dominica

Long Name: Commonwealth of Dominica

Location: Caribbean, island between the Caribbean Sea and the North Atlantic Ocean, about one-half of the way from Puerto Rico to Trinidad and Tobago

Capital: Roseau

Government Type: parliamentary democracy

Currency: EC dollar (EC$)

Area: 290 sq. mi. (750 sq. km)

Population: 64,881

Population Growth Rate: -1.3%

Birth Rate (per 1,000): 17

Death Rate (per 1,000): 6

337

Life Expectancy (from birth): 77.8

Total Fertility Rate (average children per woman): 1.9

Religion: Roman Catholic 77%, Protestant 15%

Languages: English (official), French patois

Literacy: 94%

GDP Per Capita: $2,500

Occupations: agriculture 40%, industry and commerce 32%, services 28%

Climate: tropical; moderated by northeast trade winds; heavy rainfall

Terrain: rugged mountains of volcanic origin

Dominican Republic

Long Name: Dominican Republic

Location: Caribbean, eastern two-thirds of the island of Hispaniola, between the Caribbean Sea and the North Atlantic Ocean, east of Haiti

Capital: Santo Domingo

Government Type: republic

Currency: Dominican peso (RD$)

Area: 18,675 sq. mi. (48,380 sq. km)

Population: 8,130,000

Population Growth Rate: 1.6%

Birth Rate (per 1,000): 26

Death Rate (per 1,000): 6

Life Expectancy (from birth): 69.7

Total Fertility Rate (average children per woman): 3.1

Religion: Roman Catholic 95%

Languages: Spanish

Literacy: 82.1%

GDP Per Capita: $3,670

Occupations: agriculture 50%, services and government 32%, industry 18%

Climate: tropical maritime; little seasonal temperature variation; seasonal variation in rainfall

Terrain: rugged highlands and mountains interspersed with fertile valleys

Ecuador

Long Name: Republic of Ecuador

Location: Western South America, bordering the Pacific Ocean at the Equator, between Colombia and Peru

Capital: Quito

Government Type: republic

Currency: sucre (S/)

Area: 106,860 sq. mi. (276,840 sq. km)

Population: 12,562,000

Population Growth Rate: 1.9%

Birth Rate (per 1,000): 23

Death Rate (per 1,000): 5

Life Expectancy (from birth): 71.8

Total Fertility Rate (average children per woman): 2.8

Religion: Roman Catholic 95%

Languages: Spanish (official), Amerindian languages (especially Quechua)

Literacy: 90.1%

GDP Per Capita: $4,100

Occupations: agriculture 29%, manufacturing 18%, commerce 15%, services and other activities 38%

Climate: tropical along coast becoming cooler inland

Terrain: coastal plain (costa); inter-Andean central highlands (sierra); flat to rolling eastern jungle (oriente)

Egypt

Long Name: Arab Republic of Egypt

Location: Northern Africa, bordering the Mediterranean Sea, between Libya and Israel

Capital: Cairo

Government Type: republic

Currency: Egyptian pound (£E)

Area: 384,244 sq. mi. (995,450 sq. km)

Population: 67,274,000

Population Growth Rate: 1.9%

Birth Rate (per 1,000): 27

Death Rate (per 1,000): 8

Life Expectancy (from birth): 62.1

Total Fertility Rate (average children per woman): 3.4

Religion: Muslim (mostly Sunni) 94%, Coptic Christian and other 6%

Languages: Arabic (official), English and French widely understood by educated classes

Literacy: 51.4%

GDP Per Capita: $2,900

Occupations: agriculture 40%, services, including government 38%, industry 22%

Climate: desert; hot, dry summers with moderate winters

Terrain: vast desert plateau interrupted by Nile valley and delta

El Salvador

Long Name: Republic of El Salvador

Location: Central America, bordering the North Pacific Ocean, between Guatemala and Honduras

Capital: San Salvador

Government Type: republic

Currency: Salvadoran colon (C)

Area: 7,998 sq. mi. (20,720 sq. km)

Population: 5,839,000

Population Growth Rate: 1.6%

Birth Rate (per 1,000): 27

Death Rate (per 1,000): 6

Life Expectancy (from birth): 69.7

Total Fertility Rate (average children per woman): 3.1

Religion: Roman Catholic 75%

Languages: Spanish, Nahua (among some Amerindians)

Literacy: 71.5%

GDP Per Capita: $2,080

Occupations: agriculture 40%, commerce 16%, manufacturing 15%, government 13%, financial services 9%, transportation 6%, other 1%

Climate: tropical; rainy season (May to October); dry season (November to April)

Terrain: mostly mountains; narrow coastal belt and central plateau

Equatorial Guinea

Long Name: Republic of Equatorial Guinea

Location: Western Africa, bordering the Bight of Biafra, between Cameroon and Gabon

Capital: Malabo

Government Type: republic in transition to multiparty democracy

Currency: Communaute Financiere Africaine franc (CFAF)

Area: 10,827 sq. mi. (28,050 sq. km)

Population: 465,746

Population Growth Rate: 2.6

Birth Rate (per 1,000): 39

Death Rate (per 1,000): 13

Life Expectancy (from birth): 53.9

Total Fertility Rate (average children per woman): 5.1

Religion: nominally Christian and predominantly Roman Catholic, pagan practices

Languages: Spanish (official), pidgin English, Fang, Bubi, Ibo

Literacy: 78.5%

GDP Per Capita: $800

Occupations: agriculture, industry, services

Climate: tropical; always hot, humid

Terrain: coastal plains rise to interior hills; islands are volcanic

Eritrea

Long Name: State of Eritrea

Location: Eastern Africa, bordering the Red Sea, between Djibouti and Sudan

Capital: Asmara

Government Type: transitional government

Currency: nakfa

Area: 46,830 sq. mi. (121,320 sq. km)

Population: 3,985,000

Population Growth Rate: 3.4%

Birth Rate (per 1,000): 43

Death Rate (per 1,000): 13

Life Expectancy (from birth): 55.3

Total Fertility Rate (average children per woman): 6.0

341

Religion: Muslim, Coptic Christian, Roman Catholic, Protestant

Languages: Afar, Amharic, Arabic, Italian, Tigre and Kunama, Tigrinya, minor tribal languages

GDP Per Capita: $570

Occupations: agriculture, services

Climate: hot, dry desert strip along Red Sea coast; cooler and wetter in the central highlands (up to 61 cm of rainfall annually); semiarid in western hills and lowlands; rainfall heaviest during June-September except on coastal desert

Terrain: dominated by extension of Ethiopian north-south trending highlands, descending on the east to a coastal desert plain, on the northwest to hilly climate, and on the southwest to flat-to-rolling plains

Estonia

Long Name: Republic of Estonia

Location: Eastern Europe, bordering the Baltic Sea and Gulf of Finland, between Latvia and Russia

Capital: Tallinn

Government Type: republic

Currency: Estonian kroon (EEK)

Area: 16,679 sq. mi. (43,211 sq. km)

Population: 1,409,000

Population Growth Rate: -1%

Birth Rate (per 1,000): 9

Death Rate (per 1,000): 14

Life Expectancy (from birth): 68.5

Total Fertility Rate (average children per woman): 1.3

Religion: Evangelical Lutheran; others include Baptist, Methodist, 7th Day Adventist, Roman Catholic, Pentecostal, Word of Life, Seventh Day Baptist, Judaism

Languages: Estonian (official), Latvian, Lithuanian, Russian, other

Literacy: 100%

GDP Per Capita: $5,560

Occupations: industry and construction 42%, agriculture and forestry 20%, other 38%

Climate: maritime; wet, moderate winters; cool summers

342   **Terrain:** marshy, lowlands

Ethiopia

Long Name: Federal Democratic Republic of Ethiopia

Location: Eastern Africa, west of Somalia

Capital: Addis Ababa

Government Type: federal republic

Currency: birr (Br)

Area: 432,198 sq. mi. (1,119,683 sq. km)

Population: 59,680,000

Population Growth Rate: 2.2%

Birth Rate (per 1,000): 45

Death Rate (per 1,000): 21

Life Expectancy (from birth): 40.9

Total Fertility Rate (average children per woman): 6.9

Religion: Muslim 45%–50%, Ethiopian Orthodox 35%–40%, animist 12%, other 3%–8%

Languages: Amharic (official), Tigrinya, Orominga, Guaraginga, Somali, Arabic, English (major foreign language taught in schools)

Literacy: 35.5%

GDP Per Capita: $430

Occupations: agriculture and animal husbandry 80%, government and services 12%, industry and construction 8%

Climate: tropical monsoon with wide topographic-induced variation

Terrain: high plateau with central mountain range divided by Great Rift Valley

Fiji

Long Name: Republic of Fiji

Location: Oceania, island group in the South Pacific Ocean, about two-thirds of the way from Hawaii to New Zealand

Capital: Suva

Government Type: republic

Currency: Fijian dollar (F$)

Area: 7,052 sq. mi. (18,270 sq. km)

Population: 812,918

Population Growth Rate: 1.3%

Birth Rate (per 1,000): 23

Death Rate (per 1,000): 6

Life Expectancy (from birth): 66.3

Total Fertility Rate (average children per woman): 2.7

Religion: Christian 52% (Methodist 37%, Roman Catholic 9%), Hindu 38%, Muslim 8%, other 2%

Languages: English (official), Fijian, Hindustani

Literacy: 91.6%

GDP Per Capita: $6,500

Occupations: subsistence agriculture 67%, wage earners 18%, salary earners 15%

Climate: tropical marine; only slight seasonal temperature variation

Terrain: mostly mountains of volcanic origin

Finland

Long Name: Republic of Finland

Location: Northern Europe, bordering the Baltic Sea, Gulf of Bothnia, and Gulf of Finland, between Sweden and Russia

Capital: Helsinki

Government Type: republic

Currency: markka (Finmark)

Area: 117,911 sq. mi. (305,470 sq. km)

Population: 5,158,000

Population Growth Rate: 0.2%

Birth Rate (per 1,000): 11

Death Rate (per 1,000): 10

Life Expectancy (from birth): 77.2

Total Fertility Rate (average children per woman): 1.7

Religion: Evangelical Lutheran 89%, Greek Orthodox 1%, none 9%, other 1%

Languages: Finnish 93.5% (official), Swedish 6.3% (official), small Lapp- and Russian-speaking minorities

Literacy: 100%

GDP Per Capita: $19,000

Occupations: public services 30.4%, industry 20.9%, commerce 15%, finance, insurance, and business services 10.2%, agriculture and forestry 8.6%, transport and communications 7.7%, construction 7.2%

Climate: cold temperate; potentially subarctic, but comparatively mild because of moderating influence of the North Atlantic Current, Baltic Sea, and more than 60,000 lakes

Terrain: mostly low, flat to rolling plains interspersed with lakes and low hills

France

Long Name: French Republic

Location: Western Europe, bordering the Bay of Biscay and English Channel, between Belgium and Spain southeast of the UK, bordering the Mediterranean Sea, between Italy and Spain

Capital: Paris

Government Type: republic

Currency: French franc (F)

Area: 210,613 sq. mi. (545,630 sq. km)

Population: 58,978,000

Population Growth Rate: 0.3%

Birth Rate (per 1,000): 12

Death Rate (per 1,000): 9

Life Expectancy (from birth): 78.5

Total Fertility Rate (average children per woman): 1.6

Religion: Roman Catholic 90%, Protestant 2%, Jewish 1%, Muslim (North African workers) 1%, unaffiliated 6%

Languages: French 100%, rapidly declining regional dialects and languages (Provencal, Breton, Alsatian, Corsican, Catalan, Basque, Flemish)

Literacy: 99%

GDP Per Capita: $20,900

Occupations: services 69%, industry 26%, agriculture 5%

Climate: generally cool winters and mild summers; mild winters and hot summers along the Mediterranean

Terrain: mostly flat plains or gently rolling hills in north and west; remainder is mountainous, especially Pyrenees in south, Alps in east

Gabon

Long Name: Gabonese Republic

Location: Western Africa, bordering the Atlantic Ocean at the Equator, between Republic of the Congo and Equatorial Guinea

Capital: Libreville

345

Government Type: republic; multiparty presidential regime

Currency: Communaute Financiere Africaine franc (CFAF)

Area: 99,461 sq. mi. (257,670 sq. km)

Population: 1,226,000

Population Growth Rate: 1.5%

Birth Rate (per 1,000): 28

Death Rate (per 1,000): 13

Life Expectancy (from birth): 56.5

Total Fertility Rate (average children per woman): 3.8

Religion: Christian 55%–75%, Muslim less than 1%, animist

Languages: French (official), Fang, Myene, Bateke, Bapounou/Eschira, Bandjabi

Literacy: 63.2%

GDP Per Capita: $5,400

Occupations: agriculture 65%, industry and commerce, services

Climate: tropical; always hot, humid

Terrain: narrow coastal plain; hilly interior; savanna in east and south

The Gambia

Long Name: Republic of The Gambia

Location: Western Africa, bordering the North Atlantic Ocean and Senegal

Capital: Banjul

Government Type: republic under multiparty democratic rule

Currency: dalasi (D)

Area: 3,860 sq. mi. (10,000 sq. km)

Population: 1,336,000

Population Growth Rate: 3.4%

Birth Rate (per 1,000): 43

Death Rate (per 1,000): 13

Life Expectancy (from birth): 53.9

Total Fertility Rate (average children per woman): 5.9

Religion: Muslim 90%, Christian 9%, indigenous beliefs 1%

Languages: English (official), Mandinka, Wolof, Fula, other indigenous vernaculars

Literacy: 38.6%

GDP Per Capita: $1,100

Occupations: agriculture 75%, industry, commerce, and services 18.9%, government 6.1%

Climate: tropical; hot, rainy season (June to November); cooler, dry season (November to May)

Terrain: flood plain of the Gambia River flanked by some low hills

Georgia

Long Name: Georgia

Location: Southwestern Asia, bordering the Black Sea, between Turkey and Russia

Capital: T'bilisi

Government Type: republic

Currency: lari

Area: 26,904 sq. mi. (69,700 sq. km)

Population: 5,066,000

Population Growth Rate: -0.9%

Birth Rate (per 1,000): 12

Death Rate (per 1,000): 14

Life Expectancy (from birth): 64.8

Total Fertility Rate (average children per woman): 1.5

Religion: Christian Orthodox 75% (Georgian Orthodox 65%, Russian Orthodox 10%), Muslim 11%, Armenian Apostolic 8%, unknown 6%

Languages: Georgian 71% (official), Russian 9%, Armenian 7%, Azeri 6%, other 7%

Literacy: 99%

GDP Per Capita: $1,350

Occupations: industry and construction 31%, agriculture and forestry 25%, other 44%

Climate: warm and pleasant; Mediterranean-like on Black Sea coast

Terrain: largely mountainous with Great Caucasus Mountains in the north and Lesser Caucasus Mountains in the south; Kolkhida Lowland opens to the Black Sea in the west; Mtkvari River Basin in the east; good soils in river valley flood plains, foothills of Kolkhida Lowland

Germany

Long Name: Federal Republic of Germany

Location: Central Europe, bordering the Baltic Sea and the North Sea, between the Netherlands and Poland, south of Denmark

Capital: Berlin

Government Type: federal republic

Currency: deutsche mark (DM)

Area: 134,915 sq. mi. (349,520 sq. km)

Population: 82,087,000

Population Growth Rate: 0%

Birth Rate (per 1,000): 9

Death Rate (per 1,000): 11

Life Expectancy (from birth): 77.0

Total Fertility Rate (average children per woman): 1.2

Religion: Protestant 38%, Roman Catholic 34%, Muslim 1.7%, unaffiliated or other 26.3%

Languages: German

Literacy: 99%

GDP Per Capita: $20,400

Occupations: services 56%, industry 41%, agriculture 3%

Climate: temperate and marine; cool, cloudy, wet winters and summers; occasional warm, tropical foehn wind; high relative humidity

Terrain: lowlands in north; uplands in center; Bavarian Alps in south

Ghana

Long Name: Republic of Ghana

Location: Western Africa, bordering the Gulf of Guinea, between Cote d'Ivoire and Togo

Capital: Accra

Government Type: constitutional democracy

Currency: new cedi (C)

Area: 88,788 sq. mi. (230,020 sq. km)

Population: 18,497,000

Population Growth Rate: 2.1%

Birth Rate (per 1,000): 33

Death Rate (per 1,000): 11

Life Expectancy (from birth): 56.8

Total Fertility Rate (average children per woman): 4.3

Religion: indigenous beliefs 38%, Muslim 30%, Christian 24%, other 8%

Languages: English (official), African languages (including Akan, Moshi-Dagomba, Ewe, and Ga)

Literacy: 64.5%

GDP Per Capita: $1,530

Occupations: agriculture and fishing 54.7%, industry 18.7%, sales and clerical 15.2%, professional 3.7%, services, transportation, and communications 7.7%

Climate: tropical; warm and comparatively dry along southeast coast; hot and humid in southwest; hot and dry in north

Terrain: mostly low plains; dissected plateau in south-central area

Greece

Long Name: Hellenic Republic

Location: Southern Europe, bordering the Aegean Sea, Ionian Sea, and the Mediterranean Sea, between Albania and Turkey

Capital: Athens

Government Type: parliamentary republic

Currency: drachma (Dr)

Area: 50,489 sq. mi. (130,800 sq. km)

Population: 10,662,000

Population Growth Rate: 0.4%

Birth Rate (per 1,000): 10

Death Rate (per 1,000): 9

Life Expectancy (from birth): 78.3

Total Fertility Rate (average children per woman): 1.3

Religion: Greek Orthodox 98%, Muslim 1.3%, other 0.7%

Languages: Greek (official), English, French

Literacy: 95%

GDP Per Capita: $10,000

Occupations: services 52%, agriculture 23%, industry 25%

Climate: temperate; mild, wet winters; hot, dry summers

Terrain: mostly mountains with ranges extending into sea as peninsulas or chains of islands

Grenada

Long Name: Grenada

Location: Caribbean, island between the Caribbean Sea and Atlantic Ocean, north of Trinidad and Tobago

Capital: Saint George's

Government Type: parliamentary democracy

Currency: EC dollar (EC$)

Area: 131 sq. mi. (340 sq. km)

Population: 96,217

Population Growth Rate: 0.8%

Birth Rate (per 1,000): 28

Death Rate (per 1,000): 5

Life Expectancy (from birth): 71.4

Total Fertility Rate (average children per woman): 3.6

Religion: Roman Catholic 53%, Anglican 13.8%, other Protestant sects 33.2%

Languages: English (official), French patois

Literacy: 98%

GDP Per Capita: $3,160

Occupations: services 31%, agriculture 24%, construction 8%, manufacturing 5%, other 32%

Climate: tropical; tempered by northeast trade winds

Terrain: volcanic in origin with central mountains

Guatemala

Long Name: Republic of Guatemala

Location: Central America, bordering the Caribbean Sea, between Honduras and Belize and bordering the North Pacific Ocean, between El Salvador and Mexico

Capital: Guatemala

Government Type: republic

Currency: quetzal (Q)

Area: 41,854 sq. mi. (108,430 sq. km)

Population: 12,008,000

Population Growth Rate: 2.7%

Birth Rate (per 1,000): 36

Death Rate (per 1,000): 7

Life Expectancy (from birth): 66.0

Total Fertility Rate (average children per woman): 4.8

Religion: Roman Catholic, Protestant, traditional Mayan

Languages: Spanish 60%, Amerindian languages 40% (23 Amerindian languages, including Quiche, Cakchiquel, Kekchi)

Literacy: 55.6%

GDP Per Capita: $3,460

Occupations: agriculture 58%, services 14%, manufacturing 14%, commerce 7%, construction 4%, transport 2.6%

Climate: tropical; hot, humid in lowlands; cooler in highlands

Terrain: mostly mountains with narrow coastal plains and rolling limestone plateau

Guinea

Long Name: Republic of Guinea

Location: Western Africa, bordering the North Atlantic Ocean, between Guinea-Bissau and Sierra Leone

Capital: Conakry

Government Type: republic

Currency: Guinean franc (FG)

Area: 94,902 sq. mi. (245,860 sq. km)

Population: 7,477,000

Population Growth Rate: 0.8%

Birth Rate (per 1,000): 41

Death Rate (per 1,000): 18

Life Expectancy (from birth): 46.0

Total Fertility Rate (average children per woman): 5.0

Religion: Muslim 85%, Christian 8%, indigenous beliefs 7%

Languages: French (official), each tribe has its own language

Literacy: 35.9%

GDP Per Capita: $950

Occupations: agriculture 80%, industry and commerce 11%, services 5.4%, civil service 3.6%

Climate: generally hot and humid; monsoon-type rainy season (June to November) with southwesterly winds; dry season (December to May) with northeasterly harmattan winds

Terrain: generally flat coastal plain; hilly to mountainous interior

Guinea-Bissau

Long Name: Republic of Guinea-Bissau

Location: Western Africa, bordering the North Atlantic Ocean, between Guinea and Senegal

Capital: Bissau

Government Type: republic

Currency: Guinea-Bissauan peso (PG)

Area: 10,808 sq. mi. (28,000 sq. km)

Population: 1,206,000

Population Growth Rate: 2.3%

Birth Rate (per 1,000): 39

Death Rate (per 1,000): 15

Life Expectancy (from birth): 49.1

Total Fertility Rate (average children per woman): 5.2

Religion: indigenous beliefs 65%, Muslim 30%, Christian 5%

Languages: Portuguese (official), Criolo, African languages

Literacy: 54.9%

GDP Per Capita: $950

Occupations: agriculture

Climate: tropical; generally hot and humid; monsoon-type rainy season (June to November) with southwesterly winds; dry season (December to May) with northeasterly harmattan winds

Terrain: mostly low coastal plain rising to savanna in east

Guyana

Long Name: Co-operative Republic of Guyana

Location: Northern South America, bordering the North Atlantic Ocean, between Suriname and Venezuela

Capital: Georgetown

Government Type: republic

Currency: Guyanese dollar (G$)

Area: 75,984 sq. mi. (196,850 sq. km)

Population: 707,954

Population Growth Rate: -0.5%

Birth Rate (per 1,000): 18

Death Rate (per 1,000): 9

Life Expectancy (from birth): 62.3

Total Fertility Rate (average children per woman): 2.1

Religion: Christian 57%, Hindu 33%, Muslim 9%, other 1%

Languages: English, Amerindian dialects

Literacy: 98.1%

GDP Per Capita: $2,490

Occupations: agriculture, mining

Climate: tropical; hot, humid, moderated by northeast trade winds; two rainy seasons (May to mid-August, mid-November to mid-January)

Terrain: mostly rolling highlands; low coastal plain; savanna in south

Haiti

Long Name: Republic of Haiti

Location: Caribbean, western one-third of the island of Hispaniola, between the Caribbean Sea and the North Atlantic Ocean, west of the Dominican Republic

Capital: Port-au-Prince

Government Type: republic

Currency: gourde (G)

Area: 10,638 sq. mi. (27,560 sq. km)

Population: 6,781,000

Population Growth Rate: 1.5%

Birth Rate (per 1,000): 33

Death Rate (per 1,000): 14

Life Expectancy (from birth): 51.4

Total Fertility Rate (average children per woman): 4.7

Religion: Roman Catholic 80% (of which an overwhelming majority also practice Voodoo), Protestant 16% (Baptist 10%, Pentecostal 4%, Adventist 1%, other 1%), none 1%, other 3%

Languages: French (official) 10%, Creole

Literacy: 45%

GDP Per Capita: $1,000

Occupations: agriculture 66%, services 25%, industry 9%

Climate: tropical; semiarid where mountains in east cut off trade winds

Terrain: mostly rough and mountainous

Honduras

Long Name: Republic of Honduras

Location: Central America, bordering the Caribbean Sea, between Guatemala and Nicaragua and bordering the North Pacific Ocean, between El Salvador and Nicaragua

Capital: Tegucigalpa

Government Type: republic

Currency: lempira (L)

Area: 43,190 sq. mi. (111,890 sq. km)

Population: 5,997,000

Population Growth Rate: 2.3%

Birth Rate (per 1,000): 32

Death Rate (per 1,000): 7

Life Expectancy (from birth): 65.0

Total Fertility Rate (average children per woman): 4.1

Religion: Roman Catholic 97%, Protestant minority

Languages: Spanish, Amerindian dialects

Literacy: 72.7%

GDP Per Capita: $2,000

Occupations: agriculture 62%, services 20%, manufacturing 9%, construction 3%, other 6%

Climate: subtropical in lowlands; temperate in mountains

Terrain: mostly mountains in interior; narrow coastal plains

Hungary

Long Name: Republic of Hungary

Location: Central Europe, northwest of Romania

Capital: Budapest

Government Type: republic

Currency: forint (Ft)

Area: 35,643 sq. mi. (92,340 sq. km)

Population: 10,208,000

Population Growth Rate: -0.2%

Birth Rate (per 1,000): 11

Death Rate (per 1,000): 13

Life Expectancy (from birth): 70.8

Total Fertility Rate (average children per woman): 1.4

Religion: Roman Catholic 67.5%, Calvinist 20%, Lutheran 5%, atheist and other 7.5%

Languages: Hungarian 98.2%, other 1.8%

Literacy: 99%

GDP Per Capita: $7,500

Occupations: services 58.7%, industry 34.7%, agriculture 6.6%

Climate: temperate; cold, cloudy, humid winters; warm summers

Terrain: mostly flat to rolling plains; hills and low mountains on the Slovakian border

Iceland

Long Name: Republic of Iceland

Location: Northern Europe, island between the Greenland Sea and the North Atlantic Ocean, northwest of the UK

Capital: Reykjavik

Government Type: constitutional republic

Currency: Icelandic krona (IKr)

Area: 38,697 sq. mi. (100,250 sq. km)

Population: 271,033

Population Growth Rate: 0.5%

Birth Rate (per 1,000): 15

Death Rate (per 1,000): 7

Life Expectancy (from birth): 78.8

Total Fertility Rate (average children per woman): 2.0

Religion: Evangelical Lutheran 96%, other Protestant and Roman Catholic 3%, none 1%

Languages: Icelandic

Literacy: 100%

GDP Per Capita: $19,800

Occupations: commerce, transportation, and services 60%, manufacturing 12.5%, fishing and fish processing 11.8%, construction 10.8%, agriculture 4%, other 0.9%

Climate: temperate; moderated by North Atlantic Current; mild, windy winters; damp, cool summers

Terrain: mostly plateau interspersed with mountain peaks, icefields; coast deeply indented by bays and fiords

India

Long Name: Republic of India

Location: Southern Asia, bordering the Arabian Sea and the Bay of Bengal, between Burma and Pakistan

Capital: New Delhi

Government Type: federal republic

Currency: Indian rupee (Re)

Area: 1,147,651 sq. mi. (2,973,190 sq. km)

Population: 984,004,000

Population Growth Rate: 1.7%

Birth Rate (per 1,000): 26

Death Rate (per 1,000): 9

Life Expectancy (from birth): 62.9

Total Fertility Rate (average children per woman): 3.2

Religion: Hindu 80%, Muslim 14%, Christian 2.4%, Sikh 2%, Buddhist 0.7%, Jains 0.5%, other 0.4%

Languages: English enjoys associate status and is the most important language for national, political, and commercial communication. Hindi is the national language and primary tongue of 30% of the people. Bengali (official), Telugu (official), Marathi (official), Tamil (official), Urdu (official), Gujarati (official), Malayalam (official), Kannada (official), Oriya (official), Punjabi (official), Assamese (official), Kashmiri (official), Sindhi (official), Sanskrit (official); Hindustani, a popular variant of Hindu/Urdu, is spoken widely throughout northern India. Twenty-four languages each spoken by a million or more persons; numerous other languages and dialects, for the most part mutually unintelligible.

Literacy: 52%

GDP Per Capita: $1,600

Occupations: agriculture 65% or more, services 4%, manufacturing and construction 3%, communications and transport 3%

Climate: varies from tropical monsoon in south to temperate in north

Terrain: upland plain (Deccan Plateau) in south; flat to rolling plain along the Ganges; deserts in west; Himalayas in north

Indonesia

Long Name: Republic of Indonesia

Location: Southeastern Asia, archipelago between the Indian Ocean and the Pacific Ocean

Capital: Jakarta

Government Type: republic

Currency: Indonesian rupiah (Rp)

Area: 705,006 sq. mi. (1,826,440 sq. km)

Population: 212,942,000

Population Growth Rate: 1.5%

Birth Rate (per 1,000): 23

Death Rate (per 1,000): 8

Life Expectancy (from birth): 62.5

Total Fertility Rate (average children per woman): 2.6

Religion: Muslim 87%, Protestant 6%, Roman Catholic 3%, Hindu 2%, Buddhist 1%, other 1%

Languages: Bahasa Indonesia (official; modified form of Malay), English, Dutch; local dialects, the most widely spoken of which is Javanese

Literacy: 83.8%

GDP Per Capita: $3,770

Occupations: agriculture 55%, manufacturing 10%, construction 4%, transport and communications 3%, other 28%

Climate: tropical; hot, humid; more moderate in highlands

Terrain: mostly coastal lowlands; larger islands have interior mountains

Iran

Long Name: Islamic Republic of Iran

Location: Middle East, bordering the Gulf of Oman, the Persian Gulf, and the Caspian Sea, between Iraq and Pakistan

Capital: Tehran

Government Type: theocratic republic

Currency: Iranian rials (IR)

Area: 631,496 sq. mi. (1,636,000 sq. km)

Population: 68,960,000

Population Growth Rate: 2%

Birth Rate (per 1,000): 31

Death Rate (per 1,000): 6

Life Expectancy (from birth): 68.3

Total Fertility Rate (average children per woman): 4.3

Religion: Shi'a Muslim 89%, Sunni Muslim 10%, Zoroastrian, Jewish, Christian, and Baha'i 1%

Languages: Persian and Persian dialects 58%, Turkic and Turkic dialects 26%, Kurdish 9%, Luri 2%, Balochi 1%, Arabic 1%, Turkish 1%, other 2%

Literacy: 72.1%

GDP Per Capita: $5,200

Occupations: agriculture 33%, manufacturing 21%

Climate: mostly arid or semiarid; subtropical along Caspian coast

Terrain: rugged, mountainous rim; high, central basin with deserts, mountains; small, discontinuous plains along both coasts

Iraq

Long Name: Republic of Iraq

Location: Middle East, bordering the Persian Gulf, between Iran and Kuwait

Capital: Baghdad

Government Type: republic

Currency: Iraqi dinar (ID)

Area: 166,815 sq. mi. (432,162 sq. km)

Population: 21,722,000

Population Growth Rate: 3.2%

Birth Rate (per 1,000): 39

Death Rate (per 1,000): 7

Life Expectancy (from birth): 66.5

Total Fertility Rate (average children per woman): 5.2

Religion: Muslim 97% (Shi'a 60%-65%, Sunni 32%-37%), Christian or other 3%

Languages: Arabic, Kurdish (official in Kurdish regions), Assyrian, Armenian

Literacy: 58%

GDP Per Capita: $2,000

Occupations: services 48%, agriculture 30%, industry 22%

Climate: mostly desert; mild to cool winters with dry, hot, cloudless summers; northern mountainous regions along Iranian and Turkish borders experience cold winters with occasionally heavy snows which melt in early spring, sometimes causing extensive flooding in central and southern Iraq

Terrain: mostly broad plains; reedy marshes along Iranian border in south; mountains along borders with Iran and Turkey

Ireland

Long Name: Ireland

Location: Western Europe, occupying five-sixths of the island of Ireland in the North Atlantic Ocean, west of Great Britain

Capital: Dublin

Government Type: republic

Currency: Irish pound (£Ir)

Area: 26,592 sq. mi. (68,890 sq. km)

Population: 3,619,000

Population Growth Rate: 0.4%

Birth Rate (per 1,000): 13

Death Rate (per 1,000): 9

Life Expectancy (from birth): 76.2

Total Fertility Rate (average children per woman): 1.8

Religion: Roman Catholic 93%, Anglican 3%, none 1%, unknown 2%, other 1%

Languages: Irish (Gaelic), spoken mainly in area located along the western seaboard, English is the language generally used

Literacy: 98%

GDP Per Capita: $16,800

Occupations: services 62.3%, manufacturing and construction 26%, agriculture, forestry, and fishing 10.6%, utilities 1.1%

Climate: temperate maritime; modified by North Atlantic Current; mild winters, cool summers; consistently humid; overcast about half the time

Terrain: mostly level to rolling interior plain surrounded by rugged hills and low mountains; sea cliffs on west coast

Israel

Long Name: State of Israel

Location: Middle East, bordering the Mediterranean Sea, between Egypt and Lebanon

Capital: Jerusalem (most countries maintain embassies in Tel Aviv)

Government Type: republic

Currency: new Israeli shekel (NIS)

Area: 7,847 sq. mi. (20,330 sq. km)

Population: 5,644,000

Population Growth Rate: 1.9%

Birth Rate (per 1,000): 20

Death Rate (per 1,000): 6

Life Expectancy (from birth): 78.4

Total Fertility Rate (average children per woman): 2.7

Religion: Judaism 82%, Islam 14% (mostly Sunni Muslim), Christian 2%, Druze and other 2%

Languages: Hebrew (official), Arabic used officially for Arab minority, English most commonly used foreign language

Literacy: 95%

GDP Per Capita: $16,400

Occupations: public services 29.3%, manufacturing 22.1%, commerce 13.9%, finance and business 10.4%, personal and other services 7.4%, construction 6.5%, transport, storage, and communications 6.3%, agriculture, forestry, and fishing 3.5%, other 0.6%

Climate: temperate; hot and dry in southern and eastern desert areas

Terrain: Negev desert in the south; low coastal plain; central mountains; Jordan Rift Valley

Italy

Long Name: Italian Republic

Location: Southern Europe, a peninsula extending into the central Mediterranean Sea, northeast of Tunisia

Capital: Rome

Government Type: republic

Currency: Italian lira (Lit)

Area: 113,492 sq. mi. (294,020 sq. km)

Population: 56,783,000

Population Growth Rate: -0.1%

Birth Rate (per 1,000): 9

Death Rate (per 1,000): 10

Life Expectancy (from birth): 78.4

Total Fertility Rate (average children per woman): 1.2

Religion: Roman Catholic 98%, other 2%

Languages: Italian, German, French, Slovene

Literacy: 97%

GDP Per Capita: $19,600

Occupations: services 61%, industry 32%, agriculture 7%

Climate: predominantly Mediterranean; Alpine in far north; hot, dry in south

Terrain: mostly rugged and mountainous; some plains, coastal lowlands

Jamaica

Long Name: Jamaica

Location: Caribbean, island in the Caribbean Sea, south of Cuba

Capital: Kingston

Government Type: parliamentary democracy

Currency: Jamaican dollar (J$)

Area: 4,180 sq. mi. (10,830 sq. km)

Population: 2,652,000

Population Growth Rate: 0.7%

Birth Rate (per 1,000): 21

Death Rate (per 1,000): 5

Life Expectancy (from birth): 75.4

Total Fertility Rate (average children per woman): 2.3

Religion: Protestant 55.9% (Church of God 18.4%, Baptist 10%, Anglican 7.1%, Seventh-Day Adventist 6.9%, Pentecostal 5.2%, Methodist 3.1%, United Church 2.7%, other 2.5%), Roman Catholic 5%, other, including some spiritual cults 39.1%

Languages: English, Creole

Literacy: 85%

GDP Per Capita: $3,260

Occupations: services 41%, agriculture 22.5%, industry 19%, unemployed 17.5%

Climate: tropical; hot, humid; temperate interior

Terrain: mostly mountainous; narrow, discontinuous coastal plain

Japan

Long Name: Japan

Location: Eastern Asia, island chain between the North Pacific Ocean and the Sea of Japan, east of the Korean Peninsula

Capital: Tokyo

Government Type: constitutional monarchy

Currency: yen (´)

Area: 144,651 sq. mi. (374,744 sq. km)

Population: 126,182,000

Population Growth Rate: 0.2%

Birth Rate (per 1,000): 10

Death Rate (per 1,000): 8

Life Expectancy (from birth): 80.0

Total Fertility Rate (average children per woman): 1.5

Religion: observe both Shinto and Buddhist 84%, other 16%

Languages: Japanese

Literacy: 99%

GDP Per Capita: $22,700

Occupations: trade and services 50%, manufacturing, mining, and construction 33%, utilities and communication 7%, agriculture, forestry, and fishing 6%, government 3%

Climate: varies from tropical in south to cool temperate in north

Terrain: mostly rugged and mountainous

Jordan

Long Name: Hashemite Kingdom of Jordan

Location: Middle East, northwest of Saudi Arabia

Capital: Amman

Government Type: constitutional monarchy

Currency: Jordanian dinar (JD)

Area: 34,309 sq. mi. (88,884 sq. km)

Population: 4,435,000

Population Growth Rate: 2.5%

Birth Rate (per 1,000): 35

Death Rate (per 1,000): 4

Life Expectancy (from birth): 72.8

Total Fertility Rate (average children per woman): 4.8

Religion: Sunni Muslim 92%, Christian 8%

Languages: Arabic (official), English

Literacy: 86.6%

GDP Per Capita: $5,000

Occupations: industry 11.4%, commerce, restaurants, and hotels 10.5%, construction 10%, transport and communications 8.7%, agriculture 7.4%, other services 52%

Climate: mostly arid desert; rainy season in west (November to April)

Terrain: mostly desert plateau in east; highland area in west; Great Rift Valley separates East and West Banks of the Jordan River

Kazakhstan

Long Name: Republic of Kazakhstan

Location: Central Asia, northwest of China

Capital: Astana (formerly called Akmola)

Government Type: republic

Currency: Kazakstani tenge

Area: 1,030,543 sq. mi. (2,669,800 sq. km)

Population: 16,847,000

Population Growth Rate: -0.2%

Birth Rate (per 1,000): 17

Death Rate (per 1,000): 10

Life Expectancy (from birth): 63.6

Total Fertility Rate (average children per woman): 2.1

Religion: Muslim 47%, Russian Orthodox 44%, Protestant 2%, other 7%

Languages: Kazak (Qazaq) official language spoken by over 40% of population, Russian official language spoken by two-thirds of population and used in everyday business

Literacy: 98%

GDP Per Capita: $2,880

Occupations: industry 27%, agriculture and forestry 23%, other 50%

Climate: continental, cold winters and hot summers, arid and semiarid

Terrain: extends from the Volga to the Altai Mountains and from the plains in western Siberia to oasis and desert in Central Asia

Kenya

Long Name: Republic of Kenya

Location: Eastern Africa, bordering the Indian Ocean, between Somalia and Tanzania

Capital: Nairobi

Government Type: republic

Currency: Kenyan shilling (KSh)

Area: 219,731 sq. mi. (569,250 sq. km)

Population: 28,337,000

Population Growth Rate: 1.7%

Birth Rate (per 1,000): 32

Death Rate (per 1,000): 14

Life Expectancy (from birth): 47.6

Total Fertility Rate (average children per woman): 4.1

Religion: Protestant (including Anglican) 38%, Roman Catholic 28%, indigenous beliefs 26%, other 8%

Languages: English (official), Swahili (official), numerous indigenous languages

Literacy: 78.1%

GDP Per Capita: $1,400

Occupations: agriculture 75%–80%, non-agriculture 20%-25%

Climate: varies from tropical along coast to arid in interior

Terrain: low plains rise to central highlands bisected by Great Rift Valley; fertile plateau in west

Kiribati

Long Name: Republic of Kiribati

Location: Oceania, group of islands in the Pacific Ocean, straddling the equator, about one-half of the way from Hawaii to Australia

Capital: Tarawa

Government Type: republic

Currency: Australian dollar ($A)

Area: 277 sq. mi. (717 sq. km)

Population: 83,976

Population Growth Rate: 1.8%

Birth Rate (per 1,000): 26

Death Rate (per 1,000): 8

Life Expectancy (from birth): 62.6

Total Fertility Rate (average children per woman): 3.1

Religion: Roman Catholic 53%, Protestant (Congregational) 41%, Seventh-Day Adventist, Baha'i, Church of God, Mormon 6%

Languages: English (official), Gilbertese

Literacy: 90%

GDP Per Capita: $800

364   **Occupations:** fishing, crafts

Climate: tropical; marine, hot, and humid, moderated by trade winds

Terrain: mostly low-lying coral atolls surrounded by extensive reefs

Kuwait

Long Name: State of Kuwait

Location: Middle East, bordering the Persian Gulf, between Iraq and Saudi Arabia

Capital: Kuwait

Government Type: nominal constitutional monarchy

Currency: Kuwaiti dinar (KD)

Area: 6,879 sq. mi. (17,820 sq. km)

Population: 1,913,000

Population Growth Rate: 4.1%

Birth Rate (per 1,000): 21

Death Rate (per 1,000): 2

Life Expectancy (from birth): 76.8

Total Fertility Rate (average children per woman): 3.4

Religion: Muslim 85% (Shi'a 30%, Sunni 45%, other 10%), Christian, Hindu, Parsi, and other 15%

Languages: Arabic (official), English widely spoken

Literacy: 78.6%

GDP Per Capita: $16,700

Occupations: government and social services 50%, industry and agriculture 25%, services 25%

Climate: dry desert; intensely hot summers; short, cool winters

Terrain: flat to slightly undulating desert plain

Kyrgyzstan

Long Name: Kyrgyz Republic

Location: Central Asia, west of China

Capital: Bishkek

Government Type: republic

Currency: Kyrgyzstani som (KGS)

Area: 73,842 sq. mi. (191,300 sq. km)

Population: 4,522,000

Population Growth Rate: 0.4%

Birth Rate (per 1,000): 22

Death Rate (per 1,000): 9

Life Expectancy (from birth): 63.8

Total Fertility Rate (average children per woman): 2.7

Religion: Muslim 75%, Russian Orthodox 20%, other 5%

Languages: Kirghiz (Kyrgyz), Russian

Literacy: 97%

GDP Per Capita: $1,290

Occupations: agriculture and forestry 40%, industry and construction 19%, other 41%

Climate: dry continental to polar in high Tien Shan; subtropical in southwest (Fergana Valley); temperate in northern foothill zone

Terrain: peaks of Tien Shan and associated valleys and basins encompass entire nation

Laos

Long Name: Lao People's Democratic Republic

Location: Southeastern Asia, northeast of Thailand

Capital: Vientiane

Government Type: Communist state

Currency: new kip (NK)

Area: 89,089 sq. mi. (230,800 sq. km)

Population: 5,261,000

Population Growth Rate: 2.8%

Birth Rate (per 1000): 41

Death Rate (per 1000): 13

Life Expectancy (from birth): 53.7

Total Fertility Rate (average children per woman): 5.7

Religion: Buddhist 60%, animist and other 40%

Languages: Lao (official), French, English, and various ethnic languages

Literacy: 56.6%

GDP Per Capita: $1,150

Occupations: agriculture 80%

Climate: tropical monsoon; rainy season (May to November); dry season (December to April)

Terrain: mostly rugged mountains; some plains and plateaus

Latvia

Long Name: Republic of Latvia

Location: Eastern Europe, bordering the Baltic Sea, between Estonia and Lithuania

Capital: Riga

Government Type: republic

Currency: Latvian lat (LVL)

Area: 24,743 sq. mi. (64,100 sq. km)

Population: 2,385,000

Population Growth Rate: -1.4%

Birth Rate (per 1000): 8

Death Rate (per 1000): 16

Life Expectancy (from birth): 67.1

Total Fertility Rate (average children per woman): 1.2

Religion: Lutheran, Roman Catholic, Russian Orthodox

Languages: Lettish (official), Lithuanian, Russian, other

Literacy: 100%

GDP Per Capita: $3,800

Occupations: industry 41%, agriculture and forestry 16%, services 43%

Climate: maritime; wet, moderate winters

Terrain: low plain

Lebanon

Long Name: Lebanese Republic

Location: Middle East, bordering the Mediterranean Sea, between Israel and Syria

Capital: Beirut

Government Type: republic

Currency: Lebanese pound (£L)

Area: 3,949 sq. mi. (10,230 sq. km)

Population: 3,506,000

Population Growth Rate: 1.6%

Birth Rate (per 1000): 23

Death Rate (per 1000): 7

Life Expectancy (from birth): 70.6

Total Fertility Rate (average children per woman): 2.3

Religion: Muslim 70% (five legally recognized Islamic groups: Alawite or Nusayri, Druze, Isma'ilite, Shi'a, Sunni), Christian 30% (11 legally recognized Christian groups: 4 Orthodox Christian, 6 Catholic, 1 Protestant)

Languages: Arabic (official), French (official), Armenian, English

Literacy: 92.4%

GDP Per Capita: $3,400

Occupations: services 60%, industry 28%, agriculture 12%

Climate: Mediterranean; mild to cool, wet winters with hot, dry summers; Lebanon mountains experience heavy winter snows

Terrain: narrow coastal plain; Al Biqa' (Bekaa Valley) separates Lebanon and Anti-Lebanon Mountains

Lesotho

Long Name: Kingdom of Lesotho

Location: Southern Africa, an enclave of South Africa

Capital: Maseru

Government Type: modified constitutional monarchy

Currency: loti (L), maloti (M)

Area: 11,715 sq. mi. (30,350 sq. km)

Population: 2,090,000

Population Growth Rate: 1.9%

Birth Rate (per 1000): 32

Death Rate (per 1000): 13

Life Expectancy (from birth): 54.0

Total Fertility Rate (average children per woman): 4.1

Religion: Christian 80%, other indigenous beliefs

Languages: Sesotho (southern Sotho), English (official), Zulu, Xhosa

Literacy: 71.3%

GDP Per Capita: $1,860

Occupations: 86% of resident population engaged in subsistence agriculture; roughly 35% of the active male wage earners work in South Africa

Climate: temperate; cool to cold, dry winters; hot, wet summers

Terrain: mostly highland with plateaus, hills, and mountains

Liberia

Long Name: Republic of Liberia

Location: Western Africa, bordering the North Atlantic Ocean, between Cote d'Ivoire and Sierra Leone

Capital: Monrovia

Government Type: republic

Currency: Liberian dollar (L$)

Area: 37,180 sq. mi. (96,320 sq. km)

Population: 2,772,000

Population Growth Rate: 5.8%

Birth Rate (per 1000): 42

Death Rate (per 1000): 11

Life Expectancy (from birth): 59.5

Total Fertility Rate (average children per woman): 6.1

Religion: traditional 70%, Muslim 20%, Christian 10%

Languages: English 20% (official), about 20 other tribal languages

Literacy: 38.3%

GDP Per Capita: $1,100

Occupations: agriculture 70.5%, services 10.8%, industry and commerce 4.5%, other 14.2%

Climate: tropical; hot, humid; dry winters with hot days and cool to cold nights; wet, cloudy summers with frequent heavy showers

Terrain: mostly flat to rolling coastal plains rising to rolling plateau and low mountains in northeast

Libya

Long Name: Socialist People's Libyan Arab Jamahiriya

Location: Northern Africa, bordering the Mediterranean Sea, between Egypt and Tunisia

Capital: Tripoli

Government Type: military dictatorship

Currency: Libyan dinar (LD)

Area: 679,182 sq. mi. (1,759,540 sq. km)

Population: 5,691,000

Population Growth Rate: 3.7%

Birth Rate (per 1000): 44

Death Rate (per 1000): 7

Life Expectancy (from birth): 65.4

369

Total Fertility Rate (average children per woman): 6.2

Religion: Sunni Muslim 97%

Languages: Arabic, Italian, English

Literacy: 76.2%

GDP Per Capita: $6,570

Occupations: industry 31%, services 27%, government 24%, agriculture 18%

Climate: Mediterranean along coast; dry, extreme desert interior

Terrain: mostly barren, flat to undulating plains, plateaus, depressions

Liechtenstein

Long Name: Principality of Liechtenstein

Location: Central Europe, between Austria and Switzerland

Capital: Vaduz

Government Type: hereditary constitutional monarchy

Currency: Swiss franc (SwF)

Area: 62 sq. mi. (160 sq. km)

Population: 31,717

Population Growth Rate: 1.1%

Birth Rate (per 1000): 13

Death Rate (per 1000): 7

Life Expectancy (from birth): 78.0

Total Fertility Rate (average children per woman): 1.6

Religion: Roman Catholic 80%, Protestant 6.9%, unknown 5.6%, other 7.5%

Languages: German (official), Alemannic dialect

Literacy: 100%

GDP Per Capita: $23,000

Occupations: services 53%, industry, trade, and building 45%, agriculture, fishing, forestry, and horticulture 2%

Climate: continental; cold, cloudy winters with frequent snow or rain; cool to moderately warm, cloudy, humid summers

Terrain: mostly mountainous (Alps) with Rhine Valley in western third

Lithuania

Long Name: Republic of Lithuania

Location: Eastern Europe, bordering the Baltic Sea, between Latvia and Russia

Capital: Vilnius

Government Type: independent, democratic republic

Currency: Lithuanian litas

Area: 25,167 sq. mi. (65,200 sq. km)

Population: 3,600,000

Population Growth Rate: -0.4%

Birth Rate (per 1000): 11

Death Rate (per 1000): 13

Life Expectancy (from birth): 68.8

Total Fertility Rate (average children per woman): 1.5

Religion: primarily Roman Catholic, others include Lutheran, Russian Orthodox, Protestant, evangelical Christian Baptist, Islam, Judaism

Languages: Lithuanian (official), Polish, Russian

Literacy: 98%

GDP Per Capita: $3,870

Occupations: industry and construction 42%, agriculture and forestry 18%, other 40%

Climate: transitional, between maritime and continental; wet, moderate winters and summers

Terrain: lowland, many scattered small lakes, fertile soil

Luxembourg

Long Name: Grand Duchy of Luxembourg

Location: Western Europe, between France and Germany

Capital: Luxembourg

Government Type: constitutional monarchy

Currency: Luxembourg franc (LuxF)

Area: 998 sq. mi. (2,586 sq. km)

Population: 425,017

Population Growth Rate: 1%

Birth Rate (per 1000): 11

Death Rate (per 1000): 9

Life Expectancy (from birth): 77.5

Total Fertility Rate (average children per woman): 1.6

Religion: Roman Catholic 97%, Protestant and Jewish 3%

Languages: Luxembourgish, German, French, English

Literacy: 100%

GDP Per Capita: $24,500

Occupations: trade, restaurants, hotels 20%; mining, quarrying, manufacturing 16%; other market services 18%; community, social, personal services 14%; construction 11%; finance, insurance, real estate, business services 9%; transport, storage, communications 8%; agriculture, hunting, forestry, fishing 1%; electricity, gas, water 1%

Climate: modified continental with mild winters, cool summers

Terrain: mostly gently rolling uplands with broad, shallow valleys; uplands to slightly mountainous in the north; steep slope down to Moselle floodplain in the southeast

Macedonia

Long Name: The Former Yugoslav Republic of Macedonia

Location: Southeastern Europe, north of Greece

Capital: Skopje

Government Type: emerging democracy

Currency: Macedonian denar (MKD)

Area: 9,594 sq. mi. (24,856 sq. km)

Population: 2,009,000

Population Growth Rate: 0.7%

Birth Rate (per 1000): 16

Death Rate (per 1000): 8

Life Expectancy (from birth): 72.8

Total Fertility Rate (average children per woman): 2.1

Religion: Eastern Orthodox 67%, Muslim 30%, other 3%

Languages: Macedonian 70%, Albanian 21%, Turkish 3%, Serbo-Croatian 3%, other 3%

Literacy: 91%

GDP Per Capita: $960

Occupations: manufacturing and mining 40%

Climate: hot, dry summers and autumns, and relatively cold winters with heavy snowfall

Terrain: mountainous territory covered with deep basins and valleys; there are three large lakes, each divided by a frontier line; country bisected by theVardar River

Madagascar

Long Name: Republic of Madagascar

Location: Southern Africa, island in the Indian Ocean, east of Mozambique

Capital: Antananarivo

Government Type: republic

Currency: Malagasy franc (FMG)

Area: 224,474 sq. mi. (581,540 sq. km)

Population: 14,463,000

Population Growth Rate: 2.8%

Birth Rate (per 1000): 42

Death Rate (per 1000): 14

Life Expectancy (from birth): 52.9

Total Fertility Rate (average children per woman): 5.8

Religion: indigenous beliefs 52%, Christian 41%, Muslim 7%

Languages: French (official), Malagasy (official)

Literacy: 80%

GDP Per Capita: $880

Occupations: agriculture, domestic service, industry, commerce, construction, service, transportation, other

Climate: tropical along coast, temperate inland, arid in south

Terrain: narrow coastal plain, high plateau and mountains in center

Malawi

Long Name: Republic of Malawi

Location: Southern Africa, east of Zambia

Capital: Lilongwe

Government Type: multiparty democracy

Currency: Malawian kwacha (MK)

Area: 36,315 sq. mi. (94,080 sq. km)

Population: 9,840,000

Population Growth Rate: 1.7%

Birth Rate (per 1000): 40

Death Rate (per 1000): 24

Life Expectancy (from birth): 36.6

Total Fertility Rate (average children per woman): 5.6

Religion: Protestant 55%, Roman Catholic 20%, Muslim 20%, traditional indigenous beliefs 5%

373

Languages: English (official), Chichewa (official), other languages important regionally

Literacy: 56.4%

GDP Per Capita: $800

Occupations: agriculture 86%, wage earners 14%

Climate: tropical; rainy season (November to May); dry season (May to November)

Terrain: narrow elongated plateau with rolling plains, rounded hills, some mountains

Malaysia

Long Name: Malaysia

Location: Southeastern Asia, peninsula and northern one-third of the island of Borneo, bordering Indonesia and the South China Sea, south of Vietnam

Capital: Kuala Lumpur

Government Type: constitutional monarchy

Currency: ringgit (M$)

Area: 126,820 sq. mi. (328,550 sq. km)

Population: 20,933,000

Population Growth Rate: 2.1%

Birth Rate (per 1000): 27

Death Rate (per 1000): 5

Life Expectancy (from birth): 70.4

Total Fertility Rate (average children per woman): 3.4

Religion: Peninsular Malaysia—Muslim (Malays), Buddhist (Chinese), Hindu (Indians); Sabah—Muslim 38%, Christian 17%, other 45%; Sarawak—tribal religion 35%, Buddhist and Confucianist 24%, Muslim 20%, Christian 16%, other 5%

Languages: Peninsular Malaysia—Malay (official), English, Chinese dialects, Tamil; Sabah—English, Malay, numerous tribal dialects, Chinese (Mandarin and Hakka dialects predominate); Sarawak—English, Malay, Mandarin, numerous tribal languages

Literacy: 83.5%

GDP Per Capita: $10,750

Occupations: manufacturing 25%, agriculture, forestry, and fisheries 21%, local trade and tourism 17%, services 12%, government 11%, construction 8%

Climate: tropical; annual southwest (April to October) and northeast (October to February) monsoons

374 **Terrain:** coastal plains rising to hills and mountains

Maldives

Long Name: Republic of Maldives

Location: Southern Asia, group of atolls in the Indian Ocean, south-southwest of India

Capital: Male (Maale)

Government Type: republic

Currency: rufiyaa (Rf)

Area: 116 sq. mi. (300 sq. km)

Population: 290,211

Population Growth Rate: 3.4%

Birth Rate (per 1000): 40

Death Rate (per 1000): 6

Life Expectancy (from birth): 67.6

Total Fertility Rate (average children per woman): 5.8

Religion: Sunni Muslim

Languages: Maldivian Divehi (dialect of Sinhala, script derived from Arabic)

Literacy: 93.2%

GDP Per Capita: $1,620

Occupations: fishing industry and agriculture 25%, services 21%, manufacturing and construction 21%, trade, restaurants, and hotels 16%, transportation and communication 10%, other 7%

Climate: tropical; hot, humid; dry, northeast monsoon (November to March); rainy, southwest monsoon (June to August)

Terrain: flat, with white sandy beaches

Mali

Long Name: Republic of Mali

Location: Western Africa, southwest of Algeria

Capital: Bamako

Government Type: republic

Currency: Communaute Financiere Africaine franc (CFAF)

Area: 470,920 sq. mi. (1,220,000 sq. km)

Population: 10,109,000

Population Growth Rate: 3.2%

Birth Rate (per 1000): 50

Death Rate (per 1000): 19

Life Expectancy (from birth): 47.0

Total Fertility Rate (average children per woman): 7.0

Religion: Muslim 90%, indigenous beliefs 9%, Christian 1%

Languages: French (official), Bambara 80%, numerous African languages

Literacy: 31%

GDP Per Capita: $600

Occupations: agriculture 80%, services 19%, industry and commerce 1% (1981 est.)

Climate: subtropical to arid; hot and dry February to June; rainy, humid, and mild June to November; cool and dry November to February

Terrain: mostly flat to rolling northern plains covered by sand; savanna in south, rugged hills in northeast

Malta

Long Name: Republic of Malta

Location: Southern Europe, islands in the Mediterranean Sea, south of Sicily (Italy)

Capital: Valletta

Government Type: parliamentary democracy

Currency: Maltese lira (LM)

Area: 124 sq. mi. (320 sq. km)

Population: 379,563

Population Growth Rate: 0.6%

Birth Rate (per 1000): 12

Death Rate (per 1000): 7

Life Expectancy (from birth): 77.6

Total Fertility Rate (average children per woman): 1.7

Religion: Roman Catholic 98%

Languages: Maltese (official), English (official)

Literacy: 88%

GDP Per Capita: $12,60

Occupations: public services 37%, other services 28%, manufacturing and construction 25%, agriculture 2%

Climate: Mediterranean with mild, rainy winters and hot, dry summers

Terrain: mostly low, rocky, flat to dissected plains; many coastal cliffs

Marshall Islands

Long Name: Republic of the Marshall Islands

Location: Oceania, group of atolls and reefs in the North Pacific Ocean, about one-half of the way from Hawaii to Papua New Guinea

Capital: Majuro

Government Type: constitutional government

Currency: United States dollar (US$)

Area: 70 sq. mi. (181.3 sq. km)

Population: 63,031

Population Growth Rate: 3.8%

Birth Rate (per 1000): 45

Death Rate (per 1000): 7

Life Expectancy (from birth): 64.5

Total Fertility Rate (average children per woman): 6.7

Religion: Christian (mostly Protestant)

Languages: English (universally spoken and is the official language), two major Marshallese dialects from the Malayo-Polynesian family, Japanese

Literacy: 93%

GDP Per Capita: $1,68

Occupations: agriculture and tourism

Climate: wet season from May to November; hot and humid; islands border typhoon belt

Terrain: low coral limestone and sand islands

Mauritania

Long Name: Islamic Republic of Mauritania

Location: Northern Africa, bordering the North Atlantic Ocean, between Senegal and Western Sahara

Capital: Nouakchott

Government Type: republic

Currency: ouguiya (UM)

Area: 397,734 sq. mi. (1,030,400 sq. km)

Population: 2,511,000

Population Growth Rate: 2.5%

Birth Rate (per 1000): 44

Death Rate (per 1000): 15

Life Expectancy (from birth): 50.0

Total Fertility Rate (average children per woman): 6.4

Religion: Muslim 100%

Languages: Hasaniya Arabic (official), Pular, Soninke, Wolof (official)

Literacy: 37.7%

GDP Per Capita: $1,200

Occupations: agriculture 47%, services 29%, industry and commerce 14%, government 10%

Climate: desert; hot, dry, dusty

Terrain: mostly barren, flat plains of the Sahara; some central hills

Mauritius

Long Name: Republic of Mauritius

Location: Southern Africa, island in the Indian Ocean, east of Madagascar

Capital: Port Louis

Government Type: parliamentary democracy

Currency: Mauritian rupee (MauR)

Area: 714 sq. mi. (1,850 sq. km)

Population: 1,168,000

Population Growth Rate: 1.2%

Birth Rate (per 1000): 19

Death Rate (per 1000): 7

Life Expectancy (from birth): 70.9

Total Fertility Rate (average children per woman): 2.2

Religion: Hindu 52%, Christian 28.3% (Roman Catholic 26%, Protestant 2.3%), Muslim 16.6%, other 3.1%

Languages: English (official), Creole, French, Hindi, Urdu, Hakka, Bojpoori

Literacy: 82.9%

GDP Per Capita: $10,300

Occupations: construction and industry 37%, services 24%, agriculture and fishing 15%, trade, restaurants, hotels 14%, transportation and communication 7%, finance 3%

Climate: tropical, modified by southeast trade winds; warm, dry winter (May to November); hot, wet, humid summer (November to May)

Terrain: small coastal plain rising to discontinuous mountains encircling central plateau

Mexico

Long Name: United Mexican States

Location: North America, bordering the Caribbean Sea and the Gulf of Mexico, between Belize and the U.S. and bordering the North Pacific Ocean, between Guatemala and the U.S.

Capital: Mexico City

Government Type: federal republic operating under a centralized government

Currency: New Mexican peso (Mex$)

Area: 742,293 sq. mi. (1,923,040 sq. km)

Population: 98,553,000

Population Growth Rate: 1.8%

Birth Rate (per 1000): 25

Death Rate (per 1000): 5

Life Expectancy (from birth): 71.6

Total Fertility Rate (average children per woman): 2.9

Religion: nominally Roman Catholic 89%, Protestant 6%

Languages: Spanish, various Mayan dialects

Literacy: 89.6%

GDP Per Capita: $8,100

Occupations: services 31.7%, agriculture, forestry, hunting, and fishing 28%, commerce 14.6%, manufacturing 11.1%, construction 8.4%, transportation 4.7%, mining and quarrying 1.5%

Climate: varies from tropical to desert

Terrain: high, rugged mountains, low coastal plains, high plateaus, and desert

Micronesia

Long Name: Federated States of Micronesia

Location: Oceania, island group in the North Pacific Ocean, about three-quarters of the way from Hawaii to Indonesia

Capital: Palikir

Government Type: constitutional government

Currency: United States dollar (US$)

Area: 271 sq. mi. (702 sq. km)

Population: 129,658

Population Growth Rate: 3.3%

Birth Rate (per 1000): 28

Death Rate (per 1000): 6

Life Expectancy (from birth): 68.3

Total Fertility Rate (average children per woman): 3.9

Religion: Roman Catholic 50%, Protestant 47%, other and none 3%

Languages: English (official and common language), Trukese, Pohnpeian, Yapese, Kosrean

Literacy: 89%

GDP Per Capita: $1,700

Occupations: government 66%

Climate: tropical; heavy year-round rainfall, especially in the eastern islands; located on southern edge of the typhoon belt with occasionally severe damage

Terrain: islands vary geologically from high mountainous islands to low, coral atolls; volcanic outcroppings on Pohnpei, Kosrae, and Truk

Moldova

Long Name: Republic of Moldova

Location: Eastern Europe, northeast of Romania

Capital: Chisinau

Government Type: republic

Currency: Moldovan leu (MLD), plural lei

Area: 13,008 sq. mi. (33,700 sq. km)

Population: 4,458,000

Population Growth Rate: 0%

Birth Rate (per 1000): 14

Death Rate (per 1000): 12

Life Expectancy (from birth): 64.3

Total Fertility Rate (average children per woman): 1.9

Religion: Eastern Orthodox 98.5%, Jewish 1.5%

Languages: Moldovan (official, virtually the same as the Romanian language), Russian, Gagauz (a Turkish dialect)

Literacy: 96%

GDP Per Capita: $2,400

Occupations: agriculture 39.5%, industry 12%, other 48.5%

Climate: moderate winters, warm summers

Terrain: rolling steppe, gradual slope south to Black Sea

Monaco

Long Name: Principality of Monaco

Location: Western Europe, bordering the Mediterranean Sea, on the southern coast of France, near the border with Italy

Capital: Monaco

Government Type: constitutional monarchy

Currency: French franc (F)

Area: 0.73 sq. mi. (1.9 sq. km)

Population: 32,035

Population Growth Rate: 0.4%

Birth Rate (per 1000): 11

Death Rate (per 1000): 12

Life Expectancy (from birth): 78.4

Total Fertility Rate (average children per woman): 1.7

Religion: Roman Catholic 95%

Languages: French (official), English, Italian, Monegasque

Literacy: 99%

GDP Per Capita: $25,000

Occupations: tourism

Climate: Mediterranean with mild, wet winters and hot, dry summers

Terrain: hilly, rugged, rocky

Mongolia

Long Name: Mongolia

Location: Northern Asia, between China and Russia

Capital: Ulaanbaatar

Government Type: republic

Currency: tughrik (Tug)

Area: 604,090 sq. mi. (1,565,000 sq. km)

Population: 2,579,000

Population Growth Rate: 1.5%

Birth Rate (per 1000): 24

Death Rate (per 1000): 8

Life Expectancy (from birth): 61.5

Total Fertility Rate (average children per woman): 2.7

Religion: predominantly Tibetan Buddhist, Muslim 4%

Languages: Khalkha Mongol 90%, Turkic, Russian, Chinese

Literacy: 82.9%

GDP Per Capita: $2,060

Occupations: primarily herding and agricultural

Climate: desert; continental (large daily and seasonal temperature ranges)

Terrain: vast semi-desert and desert plains; mountains in west and southwest; Gobi Desert in southeast

Morocco

Long Name: Kingdom of Morocco

Location: Northern Africa, bordering the North Atlantic Ocean and the Mediterranean Sea, between Algeria and Western Sahara

Capital: Rabat

Government Type: constitutional monarchy

Currency: Moroccan dirham (DH)

Area: 172,272 sq. mi. (446,300 sq. km)

Population: 29,114

Population Growth Rate: 1.9%

Birth Rate (per 1000): 26

Death Rate (per 1000): 6

Life Expectancy (from birth): 68.5

Total Fertility Rate (average children per woman): 3.4

Religion: Muslim 98.7%, Christian 1.1%, Jewish 0.2%

Languages: Arabic (official); French is often the language of business, government, and diplomacy; Berber dialects

Literacy: 43.7%

GDP Per Capita: $3,260

Occupations: agriculture 50%, services 26%, industry 15%, other 9% (1985 est.)

Climate: Mediterranean, becoming more extreme in the interior

Terrain: northern coast and interior are mountainous with large areas of bordering plateaus, intermontane valleys, and rich coastal plains

Mozambique

Long Name: Republic of Mozambique

Location: Southern Africa, bordering the Mozambique Channel, between South Africa and Tanzania

Capital: Maputo

Government Type: republic

Currency: metical (Mt)

Area: 302,659 sq. mi. (784,090 sq. km)

Population: 18,641,000

Population Growth Rate: 2.6%

Birth Rate (per 1000): 44

Death Rate (per 1000): 18

Life Expectancy (from birth): 45.4

Total Fertility Rate (average children per woman): 6.0

Religion: indigenous beliefs 50%, Christian 30%, Muslim 20%

Languages: Portuguese (official), indigenous dialects

Literacy: 40.1%

GDP Per Capita: $670

Occupations: agriculture 80%

Climate: tropical to subtropical

Terrain: mostly coastal lowlands, uplands in center, high plateaus in northwest, mountains in west

Myanmar

Long Name: Union of Burma

Location: Southeastern Asia, bordering the Andaman Sea and the Bay of Bengal, between Bangladesh and Thailand

Capital: Rangoon

Government Type: military regime

Currency: kyat (K)

Area: 253,888 sq. mi. (657,740 sq. km)

Population: 47,305,000

Population Growth Rate: 1.6%

Birth Rate (per 1000): 29

Death Rate (per 1000): 13

Life Expectancy (from birth): 54.5

Total Fertility Rate (average children per woman): 3.7

Religion: Buddhist 89%, Christian 4% (Baptist 3%, Roman Catholic 1%), Muslim 4%, animist beliefs 1%, other 2%

Languages: Burmese, various ethnic languages

Literacy: 83.1%

GDP Per Capita: $1,120

Occupations: agriculture 65.2%, industry 14.3%, trade 10.1%, government 6.3%, other 4.1%

Climate: tropical monsoon; cloudy, rainy, hot, humid summers (southwest monsoon, June to September); less cloudy, scant rainfall, mild temperatures, lower humidity during winter (northeast monsoon, December to April)

Terrain: central lowlands ringed by steep, rugged highlands

Namibia

Long Name: Republic of Namibia

Location: Southern Africa, bordering the South Atlantic Ocean, between Angola and South Africa

Capital: Windhoek

Government Type: republic

Currency: Namibian dollar (N$)

Area: 318,611 sq. mi. (825,418 sq. km)

Population: 1,622,000

Population Growth Rate: 1.6%

Birth Rate (per 1000): 36

Death Rate (per 1000): 20

Life Expectancy (from birth): 41.5

Total Fertility Rate (average children per woman): 5.0

Religion: Christian 80% to 90% (Lutheran 50%, other Christian denominations 30%), native religions 10% to 20%

Languages: English 7% (official), Afrikaans is the common language of most of the population and about 60% of the white population, German 32%, indigenous languages (Oshivambo, Herero, Nama)

Literacy: 38%

GDP Per Capita: $3,700

Occupations: agriculture 49%, industry and commerce 25%, services 5%, government 18%, mining 3%

Climate: desert; hot, dry; rainfall sparse and erratic

Terrain: mostly high plateau; Namib Desert along coast; Kalahari Desert in east

Nauru

Long Name: Republic of Nauru

Location: Oceania, island in the South Pacific Ocean, south of the Marshall Islands

Capital: no official capital

Government Type: republic

Currency: Australian dollar ($A)

Area: 8.1 sq. mi. (21 sq. km)

Population: 10,501

Population Growth Rate: 1.3%

Birth Rate (per 1000): 18

Death Rate (per 1000): 5

Life Expectancy (from birth): 66.7

Total Fertility Rate (average children per woman): 2.1

Religion: Christian (Protestant 66%, Roman Catholic 33%)

Languages: Nauruan (official, a distinct Pacific Island language), English

Literacy: 99%

GDP Per Capita: $10,000

Occupations: mining phosphates, public administration, education, and transportation

Climate: tropical; monsoonal; rainy season (November to February)

Terrain: sandy beach rises to fertile ring around raised coral reefs with phosphate plateau in center

Nepal

Long Name: Kingdom of Nepal

Location: Southern Asia, between China and India

Capital: Kathmandu

Government Type: parliamentary democracy

Currency: Nepalese rupee (NR)

Area: 52,805 sq. mi. (136,800 sq. km)

Population: 23,698,000

Population Growth Rate: 2.5%

Birth Rate (per 1000): 36

Death Rate (per 1000): 10

Life Expectancy (from birth): 57.9

Total Fertility Rate (average children per woman): 4.9

Religion: Hindu 90%, Buddhist 5%, Muslim 3%, other 2%

Languages: Nepali (official), 20 other languages divided into numerous dialects

Literacy: 27.5%

GDP Per Capita: $1,200

Occupations: agriculture 90%, services 7%, industry 3%

Climate: varies from cool summers and severe winters in north to subtropical summers and mild winters in south

Terrain: Terai or flat river plain of the Ganges in south, central hill region, rugged Himalayas in north

Netherlands

Long Name: Kingdom of the Netherlands

Location: Western Europe, bordering the North Sea, between Belgium and Germany

Capital: Amsterdam

Government Type: constitutional monarchy

Currency: Netherlands guilder (f.)

Area: 13,093 sq. mi. (33,920 sq. km)

Population: 15,731,000

Population Growth Rate: 0.5%

Birth Rate (per 1000): 12

Death Rate (per 1000): 9

Life Expectancy (from birth): 78.0

Total Fertility Rate (average children per woman): 1.5

Religion: Roman Catholic 34%, Protestant 25%, Muslim 3%, other 2%, unaffiliated 36%

Languages: Dutch

Literacy: 99%

GDP Per Capita: $20,500

Occupations: services 73%, manufacturing and construction 23%, agriculture 4%

Climate: temperate; marine; cool summers and mild winters

Terrain: mostly coastal lowland and reclaimed land (polders); some hills in southeast

New Zealand

Long Name: New Zealand

Location: Oceania, islands in the South Pacific Ocean, southeast of Australia

Capital: Wellington

Government Type: parliamentary democracy

Currency: New Zealand dollar (NZ$)

Area: 103,707 sq. mi. (268,670 sq. km)

Population: 3,625,000

Population Growth Rate: 1%

Birth Rate (per 1000): 15

Death Rate (per 1000): 8

Life Expectancy (from birth): 77.6

Total Fertility Rate (average children per woman): 1.9

Religion: Anglican 24%, Presbyterian 18%, Roman Catholic 15%, Methodist 5%, Baptist 2%, other Protestant 3%, unspecified or none 33%

Languages: English (official), Maori

Literacy: 99% (1980 est.)

GDP Per Capita: $18,500

Occupations: services 64.6%, industry 25%, agriculture 10.4%

Climate: temperate with sharp regional contrasts

Terrain: predominately mountainous with some large coastal plains

Nicaragua

Long Name: Republic of Nicaragua

Location: Central America, bordering both the Caribbean Sea and the North Pacific Ocean, between Costa Rica and Honduras

Capital: Managua

Government Type: republic

Currency: gold cordoba (C$)

Area: 46,418 sq. mi. (120,254 sq. km)

Population: 4,583,000

Population Growth Rate: 2.9%

Birth Rate (per 1000): 36

Death Rate (per 1000): 6

Life Expectancy (from birth): 66.6

Total Fertility Rate (average children per woman): 4.3

Religion: Roman Catholic 95%, Protestant 5%

Languages: Spanish (official)

Literacy: 65.7%

GDP Per Capita: $1,800

Occupations: services 43%, agriculture 44%, industry 13% (1986 est.)

Climate: tropical in lowlands, cooler in highlands

Terrain: extensive Atlantic coastal plains rising to central interior mountains; narrow Pacific coastal plain interrupted by volcanoes

Niger

Long Name: Republic of Niger

Location: Western Africa, southeast of Algeria

Capital: Niamey

Government Type: republic

Currency: Communaute Financiere Africaine franc (CFAF)

Area: 488,946 sq. mi. (1,266,700 sq. km)

Population: 9,672,000

Population Growth Rate: 3%

Birth Rate (per 1000): 53

Death Rate (per 1000): 23

Life Expectancy (from birth): 41.5

Total Fertility Rate (average children per woman): 7.3

Religion: Muslim 80%, Christian and indigenous beliefs

Languages: French (official), Hausa, Djerma

Literacy: 13.6%

GDP Per Capita: $640

Occupations: agriculture 90%, industry and commerce 6%, government 4%

Climate: desert; mostly hot, dry, dusty; tropical in extreme south

Terrain: predominately desert plains and sand dunes; flat to rolling plains in south; hills in north

Nigeria

Long Name: Federal Republic of Nigeria

Location: Western Africa, bordering the Gulf of Guinea, between Benin and Cameroon

Capital: Abuja

Government Type: transition from military dictatorship

Currency: naira (N)

Area: 351,557 sq. mi. (910,770 sq. km)

Population: 110,532,000

Population Growth Rate: 3%

Birth Rate (per 1000): 42

Death Rate (per 1000): 13

Life Expectancy (from birth): 53.6

Total Fertility Rate (average children per woman): 6.1

Religion: Muslim 50%, Christian 40%, indigenous beliefs 10%

Languages: English (official), Hausa, Yoruba, Ibo, Fulani

Literacy: 57.1%

GDP Per Capita: $1,380

Occupations: agriculture 54%, industry, commerce, and services 19%, government 15%

Climate: varies; equatorial in south, tropical in center, arid in north

Terrain: southern lowlands merge into central hills and plateaus; mountains in southeast, plains in north

North Korea

Long Name: Democratic People's Republic of Korea

Location: Eastern Asia, northern half of the Korean Peninsula bordering the Korea Bay and the Sea of Japan, between China and South Korea

Capital: P'yongyang

Government Type: Communist state

Currency: North Korean won (Wn)

Area: 46,478 sq. mi. (120,410 sq. km)

Population: 21,234,000

Population Growth Rate: 0%

Birth Rate (per 1000): 15

Death Rate (per 1000): 16

Life Expectancy (from birth): 51.3

Total Fertility Rate (average children per woman): 1.6

Religion: Buddhism and Confucianism, some Christianity, and syncretic Chondogyo

Languages: Korean

Literacy: 99%

GDP Per Capita: $900

Occupations: agricultural 36%, nonagricultural 64%

Climate: temperate with rainfall concentrated in summer

Terrain: mostly hills and mountains separated by deep, narrow valleys; coastal plains wide in west, discontinuous in east

Norway

Long Name: Kingdom of Norway

Location: Northern Europe, bordering the North Sea and the North Atlantic Ocean, west of Sweden

Capital: Oslo

Government Type: constitutional monarchy

Currency: Norwegian krone (NKr)

Area: 118,834 sq. mi. (307,860 sq. km)

Population: 4,420,000

Population Growth Rate: 0.4%

Birth Rate (per 1000): 13

Death Rate (per 1000): 10

Life Expectancy (from birth): 78.2

Total Fertility Rate (average children per woman): 1.8

Religion: Evangelical Lutheran 87.8% (state church), other Protestant and Roman Catholic 3.8%, none 3.2%, unknown 5.2%

Languages: Norwegian (official)

Literacy: 99%

GDP Per Capita: $26,200

Occupations: services 71%, industry 23%, agriculture, forestry, and fishing 6%

Climate: temperate along coast, modified by North Atlantic Current; colder interior; rainy year-round on west coast

Terrain: glaciated; mostly high plateaus and rugged mountains broken by fertile valleys; small, scattered plains; coastline deeply indented by fjords; arctic tundra in north

Oman

Long Name: Sultanate of Oman

Location: Middle East, bordering the Arabian Sea, Gulf of Oman, and Persian Gulf, between Yemen and United Arab Emirates

Capital: Muscat

Government Type: monarchy

Currency: Omani rial (RO)

Area: 82,010 sq. mi. (212,460 sq. km)

Population: 2,364,000

Population Growth Rate: 3.5%

Birth Rate (per 1000): 38

Death Rate (per 1000): 4

Life Expectancy (from birth): 71.0

Total Fertility Rate (average children per woman): 6.1

Religion: Ibadhi Muslim 75%, Sunni Muslim, Shi'a Muslim, Hindu

Languages: Arabic (official), English, Baluchi, Urdu, Indian dialects

Literacy: 80%

GDP Per Capita: $9,500

Occupations: agriculture 37%

Climate: dry desert; hot, humid along coast; hot, dry interior; strong southwest summer monsoon (May to September) in far south

Terrain: vast central desert plain, rugged mountains in north and south

Pakistan

Long Name: Islamic Republic of Pakistan

Location: Southern Asia, bordering the Arabian Sea, between India on the east and Iran and Afghanistan on the west

Capital: Islamabad

Government Type: federal republic

Currency: Pakistani rupee (PRe)

Area: 300,586 sq. mi. (778,720 sq. km)

Population: 135,135,000

Population Growth Rate: 2.2%

Birth Rate (per 1000): 34

Death Rate (per 1000): 11

Life Expectancy (from birth): 59.1

Total Fertility Rate (average children per woman): 4.9

Religion: Muslim 97% (Sunni 77%, Shi'a 20%), Christian, Hindu, and other 3%

Languages: Punjabi 48%, Sindhi 12%, Siraiki (a Punjabi variant) 10%, Pashtu 8%, Urdu (official) 8%, Balochi 3%, Hindko 2%, Brahui 1%, English (official), Burushaski and other 8%

Literacy: 37.8%

GDP Per Capita: $2,300

Occupations: agriculture 47%, mining and manufacturing 17%, services 17%, other 19%

Climate: mostly hot, dry desert; temperate in northwest; arctic in north

Terrain: flat Indus plain in east; mountains in north and northwest; Balochistan plateau in west

Palau

Long Name: Republic of Palau

Location: Oceania, group of islands in the North Pacific Ocean, southeast of the Philippines

Capital: Koror

Government Type: constitutional government

Currency: United States dollar (US$)

Area: 177 sq. mi. (458 sq. km)

Population: 18,110

Population Growth Rate: 2%

Birth Rate (per 1000): 21

Death Rate (per 1000): 8

Life Expectancy (from birth): 67.5

Total Fertility Rate (average children per woman): 2.6

Religion: Christian and Modekngei (indigenous to Palau)

Languages: English (official), Sonsorolese, Angaur and Japanese, Tobi, Palauan

Literacy: 92%

GDP Per Capita: $5,000

Occupations: fishing and tourism

Climate: wet season May to November; hot and humid

Terrain: varying geologically from the high, mountainous main island of Babelthuap to low, coral islands usually fringed by large barrier reefs

Panama

Long Name: Republic of Panama

Location: Central America, bordering both the Caribbean Sea and the North Pacific Ocean, between Colombia and Costa Rica

Capital: Panama City

Government Type: constitutional republic

Currency: balboa (B)

Area: 29,332 sq. mi. (75,990 sq. km)

Population: 2,736,000

Population Growth Rate: 1.6%

Birth Rate (per 1000): 22

Death Rate (per 1000): 5

Life Expectancy (from birth): 74.5

Total Fertility Rate (average children per woman): 2.6

Religion: Roman Catholic 85%, Protestant 15%

Languages: Spanish (official), English 14%

Literacy: 90.8%

GDP Per Capita: $5,300

Occupations: government and community services 31.8%, agriculture, hunting, and fishing 26.8%, commerce, restaurants, and hotels 16.4%, manufacturing and mining 9.4%, construction 3.2%, transportation and communications 6.2%, finance, insurance, and real estate 4.3%

Climate: tropical; hot, humid, cloudy; prolonged rainy season (May to January), short dry season (January to May)

Terrain: interior mostly steep, rugged mountains and dissected, upland plains; coastal areas largely plains and rolling hills

Papua New Guinea

Long Name: Independent State of Papua New Guinea

Location: Southeastern Asia, group of islands including the eastern half of the island of New Guinea between the Coral Sea and the South Pacific Ocean, east of Indonesia

Capital: Port Moresby

Government Type: parliamentary democracy

Currency: kina (K)

Area: 174,804 sq. mi. (452,860 sq. km)

Population: 4,600,000

Population Growth Rate: 2.3%

Birth Rate (per 1000): 32

Death Rate (per 1000): 10

Life Expectancy (from birth): 58.1

Total Fertility Rate (average children per woman): 4.3

Religion: Roman Catholic 22%, Lutheran 16%, Presbyterian/Methodist/London Missionary Society 8%, Anglican 5%, Evangelical Alliance 4%, Seventh-Day Adventist 1%, other Protestant sects 10%, indigenous beliefs 34%

Languages: English 2%, pidgin English widespread, Motu spoken in Papua region, 715 other indigenous languages

Literacy: 72.2%

GDP Per Capita: $2,400

Occupations: agriculture 64%

Climate: tropical; northwest monsoon (December to March), southeast monsoon (May to October); slight seasonal temperature variation

Terrain: mostly mountains with coastal lowlands and rolling foothills

Paraguay

Long Name: Republic of Paraguay

Location: Central South America, northeast of Argentina

Capital: Asuncion

Government Type: republic

Currency: guarani (G)

Area: 153,358 sq. mi. (397,300 sq. km)

Population: 5,291,000

Population Growth Rate: 2.7%

Birth Rate (per 1000): 32

Death Rate (per 1000): 5

Life Expectancy (from birth): 72.2

Total Fertility Rate (average children per woman): 4.3

Religion: Roman Catholic 90%, Mennonite and other Protestant denominations

Languages: Spanish (official), Guarani

Literacy: 92.1%

GDP Per Capita: $3,200

Occupations: agriculture 45%

Climate: subtropical; substantial rainfall in the eastern portions, becoming semiarid in the far west

Terrain: grassy plains and wooded hills east of Rio Paraguay; Gran Chaco region west of Rio Paraguay mostly low, marshy plain near the river, and dry forest and thorny scrub elsewhere

Peru

Long Name: Republic of Peru

Location: Western South America, bordering the South Pacific Ocean, between Chile and Ecuador

Capital: Lima

Government Type: republic

Currency: nuevo sol (S/.)

Area: 494,080 sq. mi. (1,280,000 sq. km)

Population: 26,111,000

Population Growth Rate: 2%

Birth Rate (per 1000): 27

Death Rate (per 1000): 6

Life Expectancy (from birth): 70.0

Total Fertility Rate (average children per woman): 3.3

Religion: Roman Catholic

Languages: Spanish (official), Quechua (official), Aymara

Literacy: 88.7%

GDP Per Capita: $3,800

Occupations: agriculture, mining and quarrying, manufacturing, construction, transport, services

Climate: varies from tropical in east to dry desert in west

Terrain: western coastal plain (costa), high and rugged Andes in center (sierra), eastern lowland jungle of Amazon Basin (selva)

Philippines

Long Name: Republic of the Philippines

395

Location: Southeastern Asia, archipelago between the Philippine Sea and the South China Sea, east of Vietnam

Capital: Manila

Government Type: republic

Currency: Philippine peso (P)

Area: 115,094 sq. mi. (298,170 sq. km)

Population: 77,726,000

Population Growth Rate: 2.1%

Birth Rate (per 1000): 28

Death Rate (per 1000): 7

Life Expectancy (from birth): 66.4

Total Fertility Rate (average children per woman): 3.5

Religion: Roman Catholic 83%, Protestant 9%, Muslim 5%, Buddhist and other 3%

Languages: Philippino (official, based on Tagalog), English (official)

Literacy: 94.6%

GDP Per Capita: $2,600

Occupations: agriculture 43.4%, services 22.6%, government services 17.9%, industry and commerce 16.1%

Climate: tropical marine; northeast monsoon (November to April); southwest monsoon (May to October)

Terrain: mostly mountains with narrow to extensive coastal lowlands

Poland

Long Name: Republic of Poland

Location: Central Europe, east of Germany

Capital: Warsaw

Government Type: democratic state

Currency: zloty (Zl)

Area: 117,541 sq. mi. (304,510 sq. km)

Population: 38,607,000

Population Growth Rate: 0%

Birth Rate (per 1000): 10

Death Rate (per 1000): 10

Life Expectancy (from birth): 72.8

Total Fertility Rate (average children per woman): 1.4

Religion: Roman Catholic 95%, Eastern Orthodox, Protestant, and other 5%

Languages: Polish

Literacy: 99%

GDP Per Capita: $6,400

Occupations: industry and construction 32%, agriculture 27.6%, trade, transport, and communications 14.7%, government and other 25.7%

Climate: temperate with cold, cloudy, moderately severe winters with frequent precipitation; mild summers with frequent showers and thundershowers

Terrain: mostly flat plain; mountains along southern border

Portugal

Long Name: Portuguese Republic

Location: Southwestern Europe, bordering the North Atlantic Ocean, west of Spain

Capital: Lisbon

Government Type: parliamentary democracy

Currency: Portuguese escudo (Esc)

Area: 35,493 sq. mi. (91,951 sq. km)

Population: 9,928,000

Population Growth Rate: -0.1%

Birth Rate (per 1000): 11

Death Rate (per 1000): 10

Life Expectancy (from birth): 75.7

Total Fertility Rate (average children per woman): 1.3

Religion: Roman Catholic 97%, Protestant denominations 1%, other 2%

Languages: Portuguese

Literacy: 85%

GDP Per Capita: $12,400

Occupations: services 54.5%, manufacturing 24.4%, agriculture, forestry, fisheries 11.2%, construction 8.3%, utilities 1%, mining 0.5%

Climate: maritime temperate; cool and rainy in north, warmer and drier in south

Terrain: mountainous north of the Tagus, rolling plains in south

Qatar

Long Name: State of Qatar

Location: Middle East, peninsula bordering the Persian Gulf and Saudi Arabia

Capital: Doha

Government Type: traditional monarchy

Currency: Qatari riyal (QR)

Area: 4,415 sq. mi. (11,437 sq. km)

Population: 697,126

Population Growth Rate: 3.8%

Birth Rate (per 1000): 17

Death Rate (per 1000): 4

Life Expectancy (from birth): 73.9

Total Fertility Rate (average children per woman): 3.5

Religion: Muslim 95%

Languages: Arabic (official), English

Literacy: 79.4%

GDP Per Capita: $21,300

Occupations: oil

Climate: desert; hot, dry; humid and sultry in summer

Terrain: mostly flat and barren desert covered with loose sand and gravel

Romania

Long Name: Romania

Location: Southeastern Europe, bordering the Black Sea, between Bulgaria and Ukraine

Capital: Bucharest

Government Type: republic

Currency: leu (L)

Area: 88,911 sq. mi. (230,340 sq. km)

Population: 22,396,000

Population Growth Rate: -0.3%

Birth Rate (per 1000): 9

Death Rate (per 1000): 12

Life Expectancy (from birth): 70.5

Total Fertility Rate (average children per woman): 1.2

Religion: Romanian Orthodox 70%, Roman Catholic 6% (of which 3% are Uniate), Protestant 6%, unaffiliated 18%

Languages: Romanian, Hungarian, German

Literacy: 97%

GDP Per Capita: $5,200

Occupations: industry 28.8%, agriculture 36.4%, other 34.8%

Climate: temperate; cold, cloudy winters with frequent snow and fog; sunny summers with frequent showers and thunderstorms

Terrain: central Transylvanian Basin is separated from the Plain of Moldavia on the east by the Carpathian Mountains and separated from the Walachian Plain on the south by the Transylvanian Alps

Russia

Long Name: Russian Federation

Location: Northern Asia (portion west of the Urals is sometimes included with Europe), bordering the Arctic Ocean, between Europe and the North Pacific Ocean

Capital: Moscow

Government Type: federation

Currency: ruble (R)

Area: 6,560,379 sq. mi. (16,995,800 sq. km)

Population: 146,861,000

Population Growth Rate: -0.3%

Birth Rate (per 1000): 10

Death Rate (per 1000): 15

Life Expectancy (from birth): 65.0

Total Fertility Rate (average children per woman): 1.3

Religion: Russian Orthodox, Muslim, other

Languages: Russian, other

Literacy: 98%

GDP Per Capita: $5,200

Occupations: diversified

Climate: ranges from steppes in the south to humid continental in much of European Russia; subarctic in Siberia to tundra climate in the polar north; winters vary from cool along Black Sea coast to frigid in Siberia; summers vary from warm in the steppes to cool along Arctic coast

Terrain: broad plain with low hills west of Urals; vast coniferous forest and tundra in Siberia; uplands and mountains along southern border regions

Rwanda

Long Name: Rwandese Republic

Location: Central Africa, east of Democratic Republic of the Congo

Capital: Kigali

Government Type: republic

Currency: Rwandan franc (RF)

Area: 9,631 sq. mi. (24,950 sq. km)

Population: 7,956,000

Population Growth Rate: 2.5%

Birth Rate (per 1000): 39

Death Rate (per 1000): 19

Life Expectancy (from birth): 41.9

Total Fertility Rate (average children per woman): 5.9

Religion: Roman Catholic 65%, Protestant 9%, Muslim 1%, indigenous beliefs and other 25%

Languages: Kinyarwanda (official), French (official), English (official), Kiswahili (Swahili) used in commercial centers

Literacy: 60.5%

GDP Per Capita: $400

Occupations: agriculture 93%, government and services 5%, industry and commerce 2%

Climate: temperate; two rainy seasons (February to April, November to January); mild in mountains with frost and snow possible

Terrain: mostly grassy uplands and hills; relief is mountainous with altitude declining from west to east

Saint Kitts and Nevis

Long Name: Federation of Saint Kitts and Nevis

Location: Caribbean, islands in the Caribbean Sea, between Puerto Rico and Trinidad and Tobago

Capital: Basseterre

Government Type: constitutional monarchy

Currency: EC dollar (EC$)

Area: 104 sq. mi. (269 sq. km)

Population: 42,291

Population Growth Rate: 1.2%

Birth Rate (per 1000): 23

Death Rate (per 1000): 9

Life Expectancy (from birth): 67.6

Total Fertility Rate (average children per woman): 2.5

Religion: Anglican, other Protestant sects, Roman Catholic

Languages: English

Literacy: 97%

GDP Per Capita: $5,700

Occupations: services 69%, manufacturing 31%

Climate: subtropical tempered by constant sea breezes; little seasonal temperature variation; rainy season (May to November)

Terrain: volcanic with mountainous interiors

Saint Lucia

Long Name: Saint Lucia

Location: Caribbean, island between the Caribbean Sea and North Atlantic Ocean, north of Trinidad and Tobago

Capital: Castries

Government Type: parliamentary democracy

Currency: EC dollar (EC$)

Area: 235 sq. mi. (610 sq. km)

Population: 152,335

Population Growth Rate: 1.1%

Birth Rate (per 1000): 22

Death Rate (per 1000): 6

Life Expectancy (from birth): 71.6

Total Fertility Rate (average children per woman): 2.4

Religion: Roman Catholic 90%, Protestant 7%, Anglican 3%

Languages: English (official), French patois

Literacy: 67%

GDP Per Capita: $4,400

Occupations: agriculture 43.4%, services 38.9%, industry and commerce 17.7% (1983 est.)

Climate: tropical, moderated by northeast trade winds; dry season from January to April, rainy season from May to August

Terrain: volcanic and mountainous with some broad, fertile valleys

Saint Vincent and the Grenadines

Long Name: Saint Vincent and the Grenadines

Location: Caribbean, islands in the Caribbean Sea, north of Trinidad and Tobago

Capital: Kingstown

Government Type: constitutional monarchy

Currency: EC dollar (EC$)

Area: 131 sq. mi. (340 sq. km)

Population: 119,818

Population Growth Rate: 0.6%

Birth Rate (per 1000): 19

Death Rate (per 1000): 5

Life Expectancy (from birth): 73.5

Total Fertility Rate (average children per woman): 2.0

Religion: Anglican, Methodist, Roman Catholic, Seventh-Day Adventist

Languages: English, French patois

Literacy: 96%

GDP Per Capita: $2,190

Occupations: agriculture 26%, industry 17%, services 57% (1980 est.)

Climate: tropical; little seasonal temperature variation; rainy season (May to November)

Terrain: volcanic, mountainous

Samoa

Long Name: Independent State of Samoa

Location: Oceania, group of islands in the South Pacific Ocean, between Hawaii and New Zealand

Capital: Apia

Government Type: constitutional monarchy

Currency: tala (WS$)

Area: 1,100 sq. mi. (2,850 sq. km)

Population: 224,713

Population Growth Rate: 2.3%

Birth Rate (per 1000): 30

Death Rate (per 1000): 6

Life Expectancy (from birth): 69.5

Total Fertility Rate (average children per woman): 3.7

Religion: Christian 99.7%

Languages: Samoan (Polynesian), English

Literacy: 97%

GDP Per Capita: $1,900

Occupations: agriculture 65%, services 30%, industry 5%

Climate: tropical; rainy season (October to March), dry season (May to October)

Terrain: narrow coastal plain with volcanic, rocky, rugged mountains in interior

San Marino

Long Name: Republic of San Marino

Location: Southern Europe, an enclave in central Italy

Capital: San Marino

Government Type: republic

Currency: Italian lire (Lit)

Area: 23 sq. mi. (60 sq. km)

Population: 24,894

Population Growth Rate: 0.7%

Birth Rate (per 1000): 11

Death Rate (per 1000): 8

Life Expectancy (from birth): 81.4

Total Fertility Rate (average children per woman): 1.5

Religion: Roman Catholic

Languages: Italian

Literacy: 96%

GDP Per Capita: $16,900

Occupations: industry 40%, agriculture 2%

Climate: Mediterranean; mild to cool winters; warm, sunny summers

Terrain: rugged mountains

Sao Tome and Principe

Long Name: Democratic Republic of Sao Tome and Principe

Location: Western Africa, island in the Gulf of Guinea, straddling the Equator, west of Gabon

403

Capital: Sao Tome

Government Type: republic

Currency: dobra (Db)

Area: 371 sq. mi. (960 sq. km)

Population: 150,123

Population Growth Rate: 3.1%

Birth Rate (per 1000): 43

Death Rate (per 1000): 8

Life Expectancy (from birth): 64.3

Total Fertility Rate (average children per woman): 6.2

Religion: Roman Catholic, Evangelical Protestant, Seventh-Day Adventist

Languages: Portuguese (official)

Literacy: 73%

GDP Per Capita: $1,000

Occupations: mostly agriculture and fishing

Climate: tropical; hot, humid; one rainy season (October to May)

Terrain: volcanic, mountainous

Saudi Arabia

Long Name: Kingdom of Saudi Arabia

Location: Middle East, bordering the Persian Gulf and the Red Sea, north of Yemen

Capital: Riyadh

Government Type: monarchy

Currency: Saudi riyal (SR)

Area: 756,785 sq. mi. (1,960,582 sq. km)

Population: 20,786,000

Population Growth Rate: 3.4%

Birth Rate (per 1000): 38

Death Rate (per 1000): 5

Life Expectancy (from birth): 70.0

Total Fertility Rate (average children per woman): 6.4

Religion: Muslim 100%

Languages: Arabic

Literacy: 62.8%

GDP Per Capita: $10,600

Occupations: government 40%, industry, construction, and oil 25%, services 30%, agriculture 5%

Climate: harsh, dry desert with great extremes of temperature

Terrain: mostly uninhabited, sandy desert

Senegal

Long Name: Republic of Senegal

Location: Western Africa, bordering the North Atlantic Ocean, between Guinea-Bissau and Mauritania

Capital: Dakar

Government Type: republic

Currency: Communaute Financiere Africaine franc (CFAF)

Area: 74,112 sq. mi. (192,000 sq. km)

Population: 9,723,000

Population Growth Rate: 3.3%

Birth Rate (per 1000): 44

Death Rate (per 1000): 11

Life Expectancy (from birth): 57.4

Total Fertility Rate (average children per woman): 6.2

Religion: Muslim 92%, indigenous beliefs 6%, Christian 2% (mostly Roman Catholic)

Languages: French (official), Wolof, Pulaar, Diola, Mandingo

Literacy: 33.1%

GDP Per Capita: $1,700

Occupations: private sector 40%, government and parapublic 60%

Climate: tropical; hot, humid; rainy season (May to November) has strong southeast winds; dry season (December to April) dominated by hot, dry, harmattan wind

Terrain: generally low, rolling plains rising to foothills in southeast

Serbia and Montenegro

Long Name: Serbia and Montenegro (Federal Republic of Yugoslavia)

Location: Southeastern Europe, bordering the Adriatic Sea, between Albania and Bosnia and Herzegovina

Capital: Belgrade

Government Type: republic

Currency: Yugoslav New Dinar (YD)

Area: 39,424 sq. mi. (102,136 sq. km)

Population: 11,224,000

Population Growth Rate: 0%

Birth Rate (per 1000): 13

Death Rate (per 1000): 9

Life Expectancy (from birth): 72.5

Total Fertility Rate (average children per woman): 1.8

Religion: Orthodox 65%, Muslim 19%, Roman Catholic 4%, Protestant 1%, other 11%

Languages: Serbo-Croatian 95%, Albanian 5%

Literacy: 98%

GDP Per Capita: $1,900

Occupations: industry 41%, services 35%, trade and tourism 12%, transportation and communication 7%, agriculture 5%

Climate: in the north, continental climate (cold winter and hot, humid summers with well-distributed rainfall); central portion, continental and Mediterranean climate; to the south, Adriatic climate along the coast, hot, dry summers and autumns and relatively cold winters with heavy snowfall inland

Terrain: extremely varied; to the north, rich fertile plains; to the east, limestone ranges and basins; to the southeast, ancient mountain and hills; to the southwest, extremely high shoreline with no islands off the coast

Seychelles

Long Name: Republic of Seychelles

Location: Eastern Africa, group of islands in the Indian Ocean, northeast of Madagascar

Capital: Victoria

Government Type: republic

Currency: Seychelles rupee (SRe)

Area: 176 sq. mi. (455 sq. km)

Population: 78,641

Population Growth Rate: 0.7%

Birth Rate (per 1000): 20

Death Rate (per 1000): 7

Life Expectancy (from birth): 70.8

Total Fertility Rate (average children per woman): 2.0

Religion: Roman Catholic 90%, Anglican 8%, other 2%

Languages: English (official), French (official), Creole

Literacy: 58%

GDP Per Capita: $6,000

Occupations: industry and commerce 31%, services 21%, government 20%, agriculture, forestry, and fishing 12%, other 16% (1985 est.)

Climate: tropical marine; humid; cooler season during southeast monsoon (late May to September); warmer season during northwest monsoon (March to May)

Terrain: Mahe Group is granitic, narrow coastal strip, rocky, hilly; others are coral, flat, elevated reefs

Sierra Leone

Long Name: Republic of Sierra Leone

Location: Western Africa, bordering the North Atlantic Ocean, between Guinea and Liberia

Capital: Freetown

Government Type: constitutional democracy

Currency: leone (Le)

Area: 27,645 sq. mi. (71,620 sq. km)

Population: 5,080,000

Population Growth Rate: 4%

Birth Rate (per 1000): 46

Death Rate (per 1000): 17

Life Expectancy (from birth): 48.6

Total Fertility Rate (average children per woman): 6.2

Religion: Muslim 60%, indigenous beliefs 30%, Christian 10%

Languages: English, Mende, Temne, Krio

Literacy: 31.4%

GDP Per Capita: $980

Occupations: agriculture 65%, industry 19%, services 16% (1981 est.)

Climate: tropical; hot, humid; summer rainy season (May to December); winter dry season (December to April)

Terrain: coastal belt of mangrove swamps, wooded hill country, upland plateau, mountains in east

Singapore

Long Name: Republic of Singapore

Location: Southeastern Asia, islands between Malaysia and Indonesia

Capital: Singapore

Government Type: republic

Currency: Singapore dollar (S$)

Area: 246 sq. mi. (637.5 sq. km)

Population: 3,490,000

Population Growth Rate: 1.2%

Birth Rate (per 1000): 14

Death Rate (per 1000): 5

Life Expectancy (from birth): 78.5

Total Fertility Rate (average children per woman): 1.5

Religion: Buddhist (Chinese), Muslim (Malays), Christian, Hindu, Sikh, Taoist, Confucianist

Languages: Chinese (official), Malay (official and national), Tamil (official), English (official)

Literacy: 91.1%

GDP Per Capita: $21,200

Occupations: financial, business, and other services 33.5%, manufacturing 25.6%, commerce 22.9%, construction 6.6%, other 11.4%

Climate: tropical; hot, humid, rainy; no pronounced rainy or dry seasons; thunderstorms occur on 40% of all days (67% of days in April)

Terrain: lowland; gently undulating central plateau contains water catchment area and nature preserve

Slovakia

Long Name: Slovak Republic

Location: Central Europe, south of Poland

Capital: Bratislava

Government Type: parliamentary democracy

Currency: koruna (Sk)

Area: 18,837 sq. mi. (48,800 sq. km)

Population: 5,393,000

Population Growth Rate: 0.1%

Birth Rate (per 1000): 10

Death Rate (per 1000): 9

Life Expectancy (from birth): 73.2

Total Fertility Rate (average children per woman): 1.3

Religion: Roman Catholic 60.3%, atheist 9.7%, Protestant 8.4%, Orthodox 4.1%, other 17.5%

Languages: Slovak (official), Hungarian

Literacy: 100%

GDP Per Capita: $8,000

Occupations: industry 29.3%, agriculture 8.9%, construction 8%, transport and communication 8.2%, services 45.6%

Climate: temperate; cool summers; cold, cloudy, humid winters

Terrain: rugged mountains in the central and northern part and lowlands in the south

Slovenia

Long Name: Republic of Slovenia

Location: Southeastern Europe, eastern Alps bordering the Adriatic Sea, between Austria and Croatia

Capital: Ljubljana

Government Type: emerging democracy

Currency: tolar (SIT)

Area: 7,819 sq. mi. (20,256 sq. km)

Population: 1,972,000

Population Growth Rate: -0.1%

Birth Rate (per 1000): 9

Death Rate (per 1000): 10

Life Expectancy (from birth): 75.2

Total Fertility Rate (average children per woman): 1.2

Religion: Roman Catholic 70.8% (including 2% Uniate), Lutheran 1%, Muslim 1%, other 27.2%

Languages: Slovenian 91%, Serbo-Croatian 6%, other 3%

Literacy: 99%

GDP Per Capita: $12,300

Occupations: agriculture and industry

Climate: Mediterranean climate on the coast, continental climate with mild to hot summers and cold winters in the plateaus and valleys to the east

Terrain: a short coastal strip on the Adriatic, an alpine mountain region adjacent to Italy, mixed mountain and valleys with numerous rivers to the east

Solomon Islands

Long Name: Solomon Islands

Location: Oceania, group of islands in the South Pacific Ocean, east of Papua New Guinea

Capital: Honiara

Government Type: parliamentary democracy

Currency: Solomon Islands dollar (SI$)

Area: 10,630 sq. mi. (27,540 sq. km)

Population: 441,039

Population Growth Rate: 3.2%

Birth Rate (per 1000): 37

Death Rate (per 1000): 4

Life Expectancy (from birth): 71.8

Total Fertility Rate (average children per woman): 5.1

Religion: Anglican 34%, Roman Catholic 19%, Baptist 17%, United (Methodist/Presbyterian) 11%, Seventh-Day Adventist 10%, other Protestant 5%, traditional beliefs 4%

Languages: Melanesian pidgin in much of the country is lingua franca, English spoken by 2% of population

Literacy: 54%

GDP Per Capita: $3,000

Occupations: services 41.5%, agriculture, forestry, and fishing 23.7%, commerce, transport, and finance 21.7%, construction, manufacturing, and mining 13.1%

Climate: tropical monsoon; few extremes of temperature and weather

Terrain: mostly rugged mountains with some low coral atolls

Somalia

Long Name: Somalia

Location: Eastern Africa, bordering the Gulf of Aden and the Indian Ocean, east of Ethiopia

Capital: Mogadishu

Government Type: transitional

Currency: Somali shilling (So. Sh.)

Area: 242,153 sq. mi. (627,340 sq. km)

Population: 6,842,000

Population Growth Rate: 4.4%

Birth Rate (per 1000): 47

Death Rate (per 1000): 19

Life Expectancy (from birth): 46.2

Total Fertility Rate (average children per woman): 7.0

Religion: Sunni Muslim

Languages: Somali (official), Arabic, Italian, English

Literacy: 24%

GDP Per Capita: $500

Occupations: agriculture (mostly pastoral nomadism) 71%, industry and services 29%

Climate: principally desert; December to February—northeast monsoon, moderate temperatures in north and very hot in south; May to October—southwest monsoon, torrid in the north and hot in the south, irregular rainfall, hot and humid periods (tangambili) between monsoons

Terrain: mostly flat to undulating plateau rising to hills in north

South Africa

Long Name: Republic of South Africa

Location: Southern Africa, at the southern tip of the continent of Africa

Capital: Pretoria (administrative), Cape Town (legislative), Bloemfontein (judicial)

Government Type: republic

Currency: rand (R)

Area: 470,886 sq. mi. (1,219,912 sq. km)

Population: 42,835,000

Population Growth Rate: 1.4%

Birth Rate (per 1000): 26

Death Rate (per 1000): 12

Life Expectancy (from birth): 55.7

Total Fertility Rate (average children per woman): 3.2

Religion: Christian 68%, Muslim 2%, Hindu 1.5%, traditional and animistic 28.5%

Languages: 11 official languages, including Afrikaans, English, Ndebele, Pedi, Sotho, Swazi, Tsonga, Tswana, Venda, Xhosa, Zulu

Literacy: 81.8%

GDP Per Capita: $5,400

Occupations: services 35%, agriculture 30%, industry 20%, mining 9%, other 6%

Climate: mostly semiarid; subtropical along east coast; sunny days, cool nights

Terrain: vast interior plateau rimmed by rugged hills and narrow coastal plain

South Korea

Long Name: Republic of Korea

Location: Eastern Asia, southern half of the Korean Peninsula bordering the Sea of Japan and the Yellow Sea, south of North Korea

Capital: Seoul

Government Type: republic

Currency: South Korean won (W)

Area: 37,901 sq. mi. (98,190 sq. km)

Population: 46,417,000

Population Growth Rate: 1%

Birth Rate (per 1000): 16

Death Rate (per 1000): 6

Life Expectancy (from birth): 74.0

Total Fertility Rate (average children per woman): 1.8

Religion: Christianity 49%, Buddhism 47%, Confucianism 3%; pervasive folk religion (shamanism), Chondogyo (religion of the Heavenly Way), and other 1%

Languages: Korean, English

Literacy: 98%

GDP Per Capita: $14,200

Occupations: services and other 52%, mining and manufacturing 27%, agriculture, fishing, forestry 21%

Climate: temperate, with rainfall heavier in summer than winter

Terrain: mostly hills and mountains; wide coastal plains in west and south

Spain

Long Name: Kingdom of Spain

Location: Southwestern Europe, bordering the Bay of Biscay, Mediterranean Sea, and North Atlantic Ocean, southwest of France

Capital: Madrid

Government Type: parliamentary monarchy

Currency: peseta (Pta)

Area: 192,768 sq. mi. (499,400 sq. km)

Population: 39,134,000

Population Growth Rate: 0.1%

Birth Rate (per 1000): 10

Death Rate (per 1000): 10

Life Expectancy (from birth): 77.6

Total Fertility Rate (average children per woman): 1.2

Religion: Roman Catholic 99%, other 1%

Languages: Castilian Spanish 74%, Catalan 17%, Galician 7%, Basque 2%

Literacy: 96%

GDP Per Capita: $15,300

Occupations: services 62%, manufacturing, mining, and construction 29%, agriculture 9%

Climate: temperate; clear, hot summers in interior, more moderate and cloudy along coast; cloudy, cold winters in interior, partly cloudy and cool along coast

Terrain: large, flat to dissected plateau surrounded by rugged hills; Pyrenees in north

Sri Lanka

Long Name: Democratic Socialist Republic of Sri Lanka

Location: Southern Asia, island in the Indian Ocean, south of India

Capital: Colombo

Government Type: republic

Currency: Sri Lankan rupee (SLRe)

Area: 24,990 sq. mi. (64,740 sq. km)

Population: 18,934,000

Population Growth Rate: 1.1%

Birth Rate (per 1000): 18

Death Rate (per 1000): 6

Life Expectancy (from birth): 72.6

Total Fertility Rate (average children per woman): 2.1

Religion: Buddhist 69%, Hindu 15%, Christian 8%, Muslim 8%

Languages: Sinhala (official and national language) 74%, Tamil (national language) 18%

Literacy: 90.2%

413

GDP Per Capita: $3,760

Occupations: agriculture 42%, services 40%, industry 18%

Climate: tropical monsoon; northeast monsoon (December to March); southwest monsoon (June to October)

Terrain: mostly low, flat to rolling plain; mountains in south-central interior

Sudan

Long Name: Republic of the Sudan

Location: Northern Africa, bordering the Red Sea, between Egypt and Eritrea

Capital: Khartoum

Government Type: transitional

Currency: Sudanese pound (£Sd)

Area: 917,136 sq. mi. (2,376,000 sq. km)

Population: 33,551,000

Population Growth Rate: 2.7%

Birth Rate (per 1000): 40

Death Rate (per 1000): 11

Life Expectancy (from birth): 56.0

Total Fertility Rate (average children per woman): 5.7

Religion: Sunni Muslim 70%, indigenous beliefs 25%, Christian 5%

Languages: Arabic (official), Nubian, Ta Bedawie, diverse dialects of Nilotic, Nilo-Hamitic, Sudanic languages, English

Literacy: 46.1%

GDP Per Capita: $860

Occupations: agriculture 80%, industry and commerce 10%, government 6%

Climate: tropical in south; arid desert in north; rainy season (April to October)

Terrain: generally flat, featureless plain; mountains in east and west

Suriname

Long Name: Republic of Suriname

Location: Northern South America, bordering the North Atlantic Ocean, between French Guiana and Guyana

Capital: Paramaribo

Government Type: republic

Currency: Surinamese guilder (Sf.)

Area: 62,327 sq. mi. (161,470 sq. km)

Population: 427,980

Population Growth Rate: 0.8%

Birth Rate (per 1000): 22

Death Rate (per 1000): 6

Life Expectancy (from birth): 70.6

Total Fertility Rate (average children per woman): 2.6

Religion: Hindu 27.4%, Roman Catholic 22.8%, Protestant 25.2%, Muslim 19.6%, indigenous beliefs 5%

Languages: Dutch (official), English, Sranang Tongo, Hindustani, Javanese

Literacy: 93%

GDP Per Capita: $3,150

Occupations: agriculture, industry, services

Climate: tropical; moderated by trade wind

Terrain: mostly rolling hills; narrow coastal plain with swamps

Swaziland

Long Name: Kingdom of Swaziland

Location: Southern Africa, between Mozambique and South Africa

Capital: Mbabane

Government Type: monarchy

Currency: lilangeni (E)

Area: 6,639 sq. mi. (17,200 sq. km)

Population: 966,462

Population Growth Rate: 2%

Birth Rate (per 1000): 41

Death Rate (per 1000): 21

Life Expectancy (from birth): 38.5

Total Fertility Rate (average children per woman): 6.0

Religion: Christian 60%, indigenous beliefs 40%

Languages: English (official, government business conducted in English), siSwati (official)

Literacy: 76.7%

GDP Per Capita: $3,800

415

Occupations: private sector 65%, public sector 35%

Climate: varies from tropical to near temperate

Terrain: mostly mountains and hills; some moderately sloping plains

Sweden

Long Name: Kingdom of Sweden

Location: Northern Europe, bordering the Baltic Sea, Gulf of Bothnia, Kattegat, and Skagerrak, between Finland and Norway

Capital: Stockholm

Government Type: constitutional monarchy

Currency: Swedish krona (SKr)

Area: 158,618 sq. mi. (410,928 sq. km)

Population: 8,887,000

Population Growth Rate: 0.3%

Birth Rate (per 1000): 12

Death Rate (per 1000): 11

Life Expectancy (from birth): 79.2

Total Fertility Rate (average children per woman): 1.8

Religion: Evangelical Lutheran 94%, Roman Catholic 1.5%, Pentecostal 1%, other 3.5%

Languages: Swedish

Literacy: 100%

GDP Per Capita: $20,800

Occupations: community, social and personal services 38.3%, mining and manufacturing 21.2%, commerce, hotels, and restaurants 14.1%, banking, insurance 9%, communications 7.2%, construction 7%, agriculture, fishing, and forestry 3.2%

Climate: temperate in south with cold, cloudy winters and cool, partly cloudy summers; subarctic in north

Terrain: mostly flat or gently rolling lowlands; mountains in west

Switzerland

Long Name: Swiss Confederation

Location: Central Europe, east of France

Capital: Bern

Government Type: federal republic

Currency: Swiss franc (SFR)

Area: 15,351 sq. mi. (39,770 sq. km)

Population: 7,260,000

Population Growth Rate: 0.2%

Birth Rate (per 1000): 11

Death Rate (per 1000): 9

Life Expectancy (from birth): 78.9

Total Fertility Rate (average children per woman): 1.5

Religion: Roman Catholic 46.7%, Protestant 40%, other 5%, no religion 8.3%

Languages: German 63.7%, French 19.2%, Italian 7.6%, Romansch 0.6%, other 8.9%

Literacy: 100%

GDP Per Capita: $22,600

Occupations: services 67%, manufacturing and construction 29%, agriculture and forestry 4%

Climate: temperate, but varies with altitude; cold, cloudy, rainy and snowy winters; cool to warm, cloudy, humid summers with occasional showers

Terrain: mostly mountains (Alps in south, Jura in northwest) with a central plateau of rolling hills, plains, and large lakes

Syria

Long Name: Syrian Arab Republic

Location: Middle East, bordering the Mediterranean Sea, between Lebanon and Turkey

Capital: Damascus

Government Type: republic under military regime

Currency: Syrian pound (£S)

Area: 71,043 sq. mi. (184,050 sq. km)

Population: 17,214,000

Population Growth Rate: 3.2%

Birth Rate (per 1000): 38

Death Rate (per 1000): 6

Life Expectancy (from birth): 67.8

Total Fertility Rate (average children per woman): 5.5

Religion: Sunni Muslim 74%, Alawite, Druze, and other Muslim sects 16%, Christian (various sects) 9%, Jewish 1%

Languages: Arabic (official), Kurdish, Armenian, Aramaic, Circassian, French

Literacy: 70.8%

GDP Per Capita: $6,300

Occupations: services 40%, agriculture 40%, industry 20%

Climate: mostly desert; hot, dry, sunny summers (June to August) and mild, rainy winters (December to February) along coast; cold weather with snow or sleet periodically hitting Damascus

Terrain: primarily semiarid and desert plateau; narrow coastal plain; mountains in west

Taiwan

Long Name: Taiwan

Location: Eastern Asia, islands bordering the East China Sea, Philippine Sea, South China Sea, and Taiwan Strait, north of the Philippines, off the southeastern coast of China

Capital: Taipei

Government Type: multiparty democratic regime

Currency: New Taiwan dollar (NT$)

Area: 12,452 sq. mi. (32,260 sq. km)

Population: 21,908,000

Population Growth Rate: 0.9%

Birth Rate (per 1000): 15

Death Rate (per 1000): 5

Life Expectancy (from birth): 76.8

Total Fertility Rate (average children per woman): 1.8

Religion: mixture of Buddhist, Confucian, and Taoist 93%, Christian 4.5%, other 2.5%

Languages: Mandarin Chinese (official), Taiwanese, Hakka dialects

Literacy: 86%

GDP Per Capita: $14,700

Occupations: services 52%, industry 38%, agriculture 10%

Climate: tropical; marine; rainy season during southwest monsoon (June to August); cloudiness is persistent and extensive all year

Terrain: eastern two-thirds mostly rugged mountains; flat to gently rolling plains in west

Tajikistan

Long Name: Republic of Tajikistan

Location: Central Asia, west of China

Capital: Dushanbe

Government Type: republic

Currency: Tajikistani ruble (TSR)

Area: 55,082 sq. mi. (142,700 sq. km)

Population: 6,020,000

Population Growth Rate: 1.3%

Birth Rate (per 1000): 28

Death Rate (per 1000): 8

Life Expectancy (from birth): 64.5

Total Fertility Rate (average children per woman): 3.5

Religion: Sunni Muslim 80%, Shi'a Muslim 5%

Languages: Tajik (official), Russian

Literacy: 98%

GDP Per Capita: $920

Occupations: agriculture and forestry 52%, manufacturing, mining, and construction 17%, services 31%

Climate: midlatitude continental, hot summers, mild winters; semiarid to polar in Pamir Mountains

Terrain: Pamirs and Alay Mountains dominate landscape; western Fergana Valley in north, Kofarnihon and Vakhsh Valleys in southwest

Tanzania

Long Name: United Republic of Tanzania

Location: Eastern Africa, bordering the Indian Ocean, between Kenya and Mozambique

Capital: Dar es Salaam

Government Type: republic

Currency: Tanzanian shilling (TSh)

Area: 342,011 sq. mi. (886,040 sq. km)

Population: 30,609,000

Population Growth Rate: 2.1%

Birth Rate (per 1000): 41

Death Rate (per 1000): 17

Life Expectancy (from birth): 46.4

Total Fertility Rate (average children per woman): 5.5

Religion: Christian 45%, Muslim 35%, indigenous beliefs 20%

Languages: Kiswahili or Swahili (official), Kiunguju, English (official), Arabic, many local languages

419

Literacy: 67.8%

GDP Per Capita: $650

Occupations: agriculture 90%, industry and commerce 10%

Climate: varies from tropical along coast to temperate in highlands

Terrain: plains along coast; central plateau; highlands in north, south

Thailand

Long Name: Kingdom of Thailand

Location: Southeastern Asia, bordering the Andaman Sea and the Gulf of Thailand, southeast of Burma

Capital: Bangkok

Government Type: constitutional monarchy

Currency: baht (B)

Area: 197,543 sq. mi. (511,770 sq. km)

Population: 60,037,000

Population Growth Rate: 1%

Birth Rate (per 1000): 17

Death Rate (per 1000): 7

Life Expectancy (from birth): 69.0

Total Fertility Rate (average children per woman): 1.8

Religion: Buddhism 95%, Islam 3.8%, Christianity 0.5%, Hinduism 0.1%, other 0.6%

Languages: Thai, English, ethnic and regional dialects

Literacy: 93.8%

GDP Per Capita: $7,700

Occupations: agriculture 57%, industry 17%, commerce 11%, services (including government) 15%

Climate: tropical; rainy, warm, cloudy southwest monsoon (mid-May to September); dry, cool northeast monsoon (November to mid-March); southern isthmus always hot and humid

Terrain: central plain; Khorat Plateau in the east; mountains elsewhere

Togo

Long Name: Togolese Republic

Location: Western Africa, bordering the Bight of Benin, between Benin and Ghana

420 **Capital:** Lome

Government Type: republic

Currency: Communaute Financiere Africaine franc (CFAF)

Area: 20,995 sq. mi. (54,390 sq. km)

Population: 4,906,000

Population Growth Rate: 3.5%

Birth Rate (per 1000): 45

Death Rate (per 1000): 10

Life Expectancy (from birth): 58.8

Total Fertility Rate (average children per woman): 6.6

Religion: indigenous beliefs 70%, Christian 20%, Muslim 10%

Languages: French (official), Ewe, Mina, Kabye, and Dagomba

Literacy: 51.7%

GDP Per Capita: $970

Occupations: agriculture 64%, industry 9%, services 21%, unemployed 6% (1981 est.)

Climate: tropical; hot, humid in south; semiarid in north

Terrain: gently rolling savanna in north; central hills; southern plateau; low coastal plain with extensive lagoons and marshes

Tonga

Long Name: Kingdom of Tonga

Location: Oceania, archipelago in the South Pacific Ocean, between Hawaii and New Zealand

Capital: Nuku'alofa

Government Type: hereditary constitutional monarchy

Currency: pa'anga (T$)

Area: 277 sq. mi. (718 sq. km)

Population: 108,207

Population Growth Rate: 0.8%

Birth Rate (per 1000): 26

Death Rate (per 1000): 6

Life Expectancy (from birth): 69.5

Total Fertility Rate (average children per woman): 3.6

Religion: Christian

Languages: Tongan, English

Literacy: 100%

GDP Per Capita: $2,140

Occupations: agriculture 70%

Climate: tropical; modified by trade winds; warm season (December to May), cool season (May to December)

Terrain: most islands have limestone base formed from uplifted coral formation; others have limestone overlying volcanic base

Trinidad and Tobago

Long Name: Republic of Trinidad and Tobago

Location: Caribbean, islands between the Caribbean Sea and the North Atlantic Ocean, northeast of Venezuela

Capital: Port-of-Spain

Government Type: parliamentary democracy

Currency: Trinidad and Tobago dollar (TT$)

Area: 1,980 sq. mi. (5,130 sq. km)

Population: 1,117,000

Population Growth Rate: -1.3%

Birth Rate (per 1000): 15

Death Rate (per 1000): 8

Life Expectancy (from birth): 70.5

Total Fertility Rate (average children per woman): 2.1

Religion: Roman Catholic 32.2%, Hindu 24.3%, Anglican 14.4%, other Protestant 14%, Muslim 6%, none or unknown 9.1%

Languages: English (official), Hindi, French, Spanish

Literacy: 97.9%

GDP Per Capita: $13,500

Occupations: construction and utilities 13%, manufacturing, mining, and quarrying 14%, agriculture 11%, services 62%

Climate: tropical; rainy season (June to December)

Terrain: mostly plains with some hills and low mountains

Tunisia

Long Name: Republic of Tunisia

Location: Northern Africa, bordering the Mediterranean Sea, between Algeria and Libya

Capital: Tunis

Government Type: republic

Currency: Tunisian dinar (TD)

Area: 59,970 sq. mi. (155,360 sq. km)

Population: 9,380,000

Population Growth Rate: 1.4%

Birth Rate (per 1000): 20

Death Rate (per 1000): 5

Life Expectancy (from birth): 73.1

Total Fertility Rate (average children per woman): 2.4

Religion: Muslim 98%, Christian 1%, Jewish 1%

Languages: Arabic (official), French

Literacy: 66.7%

GDP Per Capita: $4,800

Occupations: services 55%, industry 23%, agriculture 22%

Climate: temperate in north with mild, rainy winters and hot, dry summers; desert in south

Terrain: mountains in north; hot, dry central plain; semiarid south merges into the Sahara

Turkey

Long Name: Republic of Turkey

Location: Southwestern Asia (portion west of the Bosporus is sometimes included with Europe), bordering the Black Sea, between Bulgaria and Georgia, and bordering the Aegean Sea and the Mediterranean Sea, between Greece and Syria

Capital: Ankara

Government Type: republican parliamentary democracy

Currency: Turkish lira (TL)

Area: 297,513 sq. mi. (770,760 sq. km)

Population: 64,567,000

Population Growth Rate: 1.6%

Birth Rate (per 1000): 21

Death Rate (per 1000): 5

Life Expectancy (from birth): 72.8

Total Fertility Rate (average children per woman): 2.5

Religion: Muslim 99.8% (mostly Sunni), other (mostly Christian and Jewish) 0.2%

Languages: Turkish (official), Kurdish, Arabic

Literacy: 82.3%

GDP Per Capita: $6,100

Occupations: agriculture 47%, services 33%, industry 20%

Climate: temperate; hot, dry summers with mild, wet winters; harsher in interior

Terrain: mostly mountains; narrow coastal plain; high central plateau (Anatolia)

Turkmenistan

Long Name: Turkmenistan

Location: Central Asia, bordering the Caspian Sea, between Iran and Kazakstan

Capital: Ashgabat

Government Type: republic

Currency: Tukmen manat (TMM)

Area: 188,407 sq. mi. (488,100 sq. km)

Population: 4,298,000

Population Growth Rate: 1.6%

Birth Rate (per 1000): 26

Death Rate (per 1000): 9

Life Expectancy (from birth): 61.3

Total Fertility Rate (average children per woman): 3.3

Religion: Muslim 89%, Eastern Orthodox 9%, unknown 2%

Languages: Turkmen 72%, Russian 12%, Uzbek 9%, other 7%

Literacy: 98%

GDP Per Capita: $2,840

Occupations: agriculture and forestry 43%, industry and construction 20%, other 37%

Climate: subtropical desert

Terrain: flat and rolling sandy desert with dunes rising to mountains in the south; low mountains along border with Iran; borders Caspian Sea in west

Tuvalu

Long Name: Tuvalu

Location: Oceania, island group consisting of nine coral atolls in the South Pacific Ocean, between Hawaii and Australia

Capital: Funafuti

Government Type: democracy

Currency: Tuvaluan dollar ($T)

Area: 10 sq. mi. (26 sq. km)

Population: 10,444

Population Growth Rate: 1.4%

Birth Rate (per 1000): 23

Death Rate (per 1000): 9

Life Expectancy (from birth): 63.9

Total Fertility Rate (average children per woman): 3.1

Religion: Church of Tuvalu (Congregationalist) 97%, Seventh-Day Adventist 1.4%, Baha'i 1%, other 0.6%

Languages: Tuvaluan, English

Literacy: 95%

GDP Per Capita: $800

Occupations: fishing

Climate: tropical; moderated by easterly trade winds (March to November); westerly gales and heavy rain (November to March)

Terrain: very low-lying and narrow coral atolls

Uganda

Long Name: Republic of Uganda

Location: Eastern Africa, west of Kenya

Capital: Kampala

Government Type: republic

Currency: Ugandan shilling (USh)

Area: 77,088 sq. mi. (199,710 sq. km)

Population: 22,167,000

Population Growth Rate: 2.8%

Birth Rate (per 1000): 49

Death Rate (per 1000): 19

Life Expectancy (from birth): 42.6

Total Fertility Rate (average children per woman): 7.1

Religion: Roman Catholic 33%, Protestant 33%, Muslim 16%, indigenous beliefs 18%

Languages: English (official), Luganda, Swahili, Bantu languages, Nilotic Languages

Literacy: 61.8%

425

GDP Per Capita: $900

Occupations: agriculture 86%, industry 4%, services 10%

Climate: tropical; generally rainy with two dry seasons (December to February, June to August); semiarid in northeast

Terrain: mostly plateau with rim of mountains

Ukraine

Long Name: Ukraine

Location: Eastern Europe, bordering the Black Sea, between Poland and Russia

Capital: Kiev (Kyyiv)

Government Type: republic

Currency: hryvnia (plural hryvni)

Area: 233,028 sq. mi. (603,700 sq. km)

Population: 50,125,000

Population Growth Rate: -0.6%

Birth Rate (per 1000): 10

Death Rate (per 1000): 16

Life Expectancy (from birth): 65.8

Total Fertility Rate (average children per woman): 1.3

Religion: Ukrainian Orthodox (Moscow Patriarchate, Kiev Patriarchate), Ukrainian Autocephalous Orthodox, Ukrainian Catholic, Protestant, Jewish

Languages: Ukrainian, Russian, Romanian, Polish, Hungarian

Literacy: 98%

GDP Per Capita: $3,170

Occupations: industry and construction 33%, agriculture and forestry 21%, health, education, and culture 16%, trade and distribution 7%, transport and communication 7%, other 16%

Climate: temperate continental; Mediterranean only on the southern Crimean coast; precipitation disproportionately distributed, highest in west and north, lesser in east and southeast; winters vary from cool along the Black Sea to cold farther inland; summers are warm across the greater part of the country, hot in the south

Terrain: most of Ukraine consists of fertile plains (steppes) and plateaus, mountains being found only in the west (the Carpathians), and in the Crimean Peninsula in the extreme south

United Arab Emirates

Long Name: United Arab Emirates

Location: Middle East, bordering the Gulf of Oman and the Persian Gulf, between Oman and Saudi Arabia

Capital: Abu Dhabi

Government Type: federation

Currency: Emirian dirham (Dh)

Area: 31,992 sq. mi. (82,880 sq. km)

Population: 2,303,000

Population Growth Rate: 1.8%

Birth Rate (per 1000): 19

Death Rate (per 1000): 3

Life Expectancy (from birth): 74.9

Total Fertility Rate (average children per woman): 3.6

Religion: Muslim 96% (Shi'a 16%); Christian, Hindu, and other 4%

Languages: Arabic (official), Persian, English, Hindi, Urdu

Literacy: 79.2%

GDP Per Capita: $23,800

Occupations: industry and commerce 56%, services 38%, agriculture 6%

Climate: desert; cooler in eastern mountains

Terrain: flat, barren, coastal plain merging into rolling sand dunes of vast desert wasteland; mountains in east

United Kingdom

Long Name: United Kingdom of Great Britain and Northern Ireland

Location: Western Europe, islands including the northern one-sixth of the island of Ireland between the North Atlantic Ocean and the North Sea, northwest of France

Capital: London

Government Type: constitutional monarchy

Currency: British pound (£)

Area: 93,254 sq. mi. (241,590 sq. km)

Population: 58,970,000

Population Growth Rate: 0.3%

Birth Rate (per 1000): 12

Death Rate (per 1000): 11

Life Expectancy (from birth): 77.2

Total Fertility Rate (average children per woman): 1.7

Religion: Anglican, Roman Catholic, Muslim, Presbyterian, Methodist, Sikh, Hindu, Jewish

Languages: English, Welsh, Scottish form of Gaelic

Literacy: 100%

GDP Per Capita: $20,400

Occupations: services 62.8%, manufacturing and construction 25%, government 9.1%, energy 1.9%, agriculture 1.2%

Climate: temperate; moderated by prevailing southwest winds over the North Atlantic Current; more than 50% of the days are overcast

Terrain: mostly rugged hills and low mountains; level to rolling plains in east and southeast

United States

Long Name: United States of America

Location: North America, bordering both the North Atlantic Ocean and the North Pacific Ocean, between Canada and Mexico

Capital: Washington, DC

Government Type: federal republic

Currency: United States dollar (US$)

Area: 3,535,359 sq. mi. (9,158,960 sq. km)

Population: 270,312,000

Population Growth Rate: 0.9%

Birth Rate (per 1000): 14

Death Rate (per 1000): 9

Life Expectancy (from birth): 76.1

Total Fertility Rate (average children per woman): 2.1

Religion: Protestant 56%, Roman Catholic 28%, Jewish 2%, other 4%, none 10%

Languages: English, Spanish

Literacy: 97%

GDP Per Capita: $28,600

Occupations: managerial and professional 28.8%, technical, sales and administrative support 29.7%, services 13.6%, manufacturing, mining, transportation, and crafts 25.1%, farming, forestry, and fishing 2.8%

Climate: mostly temperate, but tropical in Hawaii and Florida and arctic in Alaska, semi-arid in the great plains west of the Mississippi River and arid in the Great Basin of the southwest; low winter temperatures in the northwest are ameliorated occasionally in January and February by warm chinook winds from the eastern slopes of the Rocky Mountains

Terrain: vast central plain, mountains in west, hills and low mountains in east; rugged mountains and broad river valleys in Alaska; rugged, volcanic topography in Hawaii

Uruguay

Long Name: Oriental Republic of Uruguay

Location: Southern South America, bordering the South Atlantic Ocean, between Argentina and Brazil

Capital: Montevideo

Government Type: republic

Currency: Uruguayan peso ($Ur)

Area: 67,017 sq. mi. (173,620 sq. km)

Population: 3,285,000

Population Growth Rate: 0.7%

Birth Rate (per 1000): 17

Death Rate (per 1000): 9

Life Expectancy (from birth): 75.5

Total Fertility Rate (average children per woman): 2.3

Religion: Roman Catholic 66%, Protestant 2%, Jewish 2%, nonprofessing or other 30%

Languages: Spanish, Portunon, Brazilero

Literacy: 97.3%

GDP Per Capita: $8,000

Occupations: government 25%, manufacturing 19%, agriculture 11%, commerce 12%, utilities, construction, transport, and communications 12%, other services 21% (1988 est.)

Climate: warm temperate; freezing temperatures almost unknown

Terrain: mostly rolling plains and low hills; fertile coastal lowland

Uzbekistan

Long Name: Republic of Uzbekistan

Location: Central Asia, north of Afghanistan

Capital: Tashkent (Toshkent)

Government Type: republic

Currency: som

Area: 164,204 sq. mi. (425,400 sq. km)

Population: 23,784,000

Population Growth Rate: 1.3%

Birth Rate (per 1000): 24

Death Rate (per 1000): 8

Life Expectancy (from birth): 64.1

Total Fertility Rate (average children per woman): 2.9

Religion: Muslim 88% (mostly Sunnis), Eastern Orthodox 9%, other 3%

Languages: Uzbek 74.3%, Russian 14.2%, Tajik 4.4%, other 7.1%

Literacy: 97%

GDP Per Capita: $2,430

Occupations: agriculture and forestry 44%, industry and construction 20%, other 36%

Climate: mostly midlatitude desert; long, hot summers, mild winters; semiarid grassland in east

Terrain: mostly flat and rolling sandy desert with dunes; broad, flat, intensely irrigated river valleys along course of Amu Darya and Syr Darya; Fergana Valley in east surrounded by mountainous Tajikistan and Kyrgyzstan; shrinking Aral Sea in west

Vanuatu

Long Name: Republic of Vanuatu

Location: Oceania, group of islands in the South Pacific Ocean, between Hawaii and Australia

Capital: Port-Vila

Government Type: republic

Currency: vatu (VT)

Area: 5,697 sq. mi. (14,760 sq. km)

Population: 185,204

Population Growth Rate: 2.1%

Birth Rate (per 1000): 29

Death Rate (per 1000): 8

Life Expectancy (from birth): 61.0

Total Fertility Rate (average children per woman): 3.7

Religion: Presbyterian 36.7%, Anglican 15%, Catholic 15%, indigenous beliefs 7.6%, Seventh-Day Adventist 6.2%, Church of Christ 3.8%, other 15.7%

Languages: English (official), French (official), pidgin (known as Bislama or Bichelama)

Literacy: 53%

GDP Per Capita: $1,230

Occupations: agriculture 65%, services 32%, industry 3%

Climate: tropical; moderated by southeast trade winds

Terrain: mostly mountains of volcanic origin; narrow coastal plains

Vatican City

Long Name: The Holy See (State of the Vatican City)

Location: Southern Europe, an enclave of Rome (Italy)

Capital: Vatican City

Government Type: monarchical-sacerdotal state

Currency: Vatican lira (VLit)

Area: 0.17 sq. mi. (0.44 sq. km)

Population: 850

Population Growth Rate: N/A

Birth Rate (per 1000): N/A

Death Rate (per 1000): N/A

Life Expectancy (from birth): N/A

Total Fertility Rate (average children per woman): N/A

Religion: Roman Catholic 100%

Languages: Italian, Latin, various other languages

Literacy: 100%

GDP Per Capita: N/A

Occupations: dignitaries, priests, nuns, guards, and 3,000 lay workers who live outside the Vatican

Climate: temperate; mild, rainy winters (September to mid-May) with hot, dry summers (May to September)

Terrain: low hill

Venezuela

Long Name: Republic of Venezuela

Location: Northern South America, bordering the Caribbean Sea and the North Atlantic Ocean, between Colombia and Guyana

Capital: Caracas

Government Type: republic

Currency: bolivar (Bs)

Area: 34,0471 sq. mi. (882,050 sq. km)

Population: 22,803,000

Population Growth Rate: 1.8%

Birth Rate (per 1000): 23

Death Rate (per 1000): 5

Life Expectancy (from birth): 72.7

Total Fertility Rate (average children per woman): 2.7

Religion: Roman Catholic 96%, Protestant 2%

Languages: Spanish (official), native dialects

Literacy: 91.1%

GDP Per Capita: $9,000

Occupations: services 64%, industry 23%, agriculture 13%

Climate: tropical; hot, humid; more moderate in highlands

Terrain: Andes Mountains and Maracaibo Lowlands in northwest; central plains (llanos); Guiana Highlands in southeast

Vietnam

Long Name: Socialist Republic of Vietnam

Location: Southeastern Asia, bordering the Gulf of Thailand, Gulf of Tonkin, and South China Sea, between China and Cambodia

Capital: Hanoi

Government Type: Communist state

Currency: new dong (D)

Area: 125,589 sq. mi. (325,360 sq. km)

Population: 76,236,000

Population Growth Rate: 1.4%

Birth Rate (per 1000): 22

Death Rate (per 1000): 7

Life Expectancy (from birth): 67.7

432   **Total Fertility Rate (average children per woman):** 2.5

Religion: Buddhist, Taoist, Roman Catholic, indigenous beliefs, Muslim, Protestant, Cao Dai, Hoa Hao

Languages: Vietnamese (official), Chinese, English, French, Khmer, tribal languages (Mon-Khmer and Malayo-Polynesian)

Literacy: 93.7%

GDP Per Capita: $1,470

Occupations: agriculture 65%, industry and services 35%

Climate: tropical in south; monsoonal in north with hot, rainy season (mid-May to mid-September) and warm, dry season (mid-October to mid-March)

Terrain: low, flat delta in south and north; central highlands; hilly, mountainous in far north and northwest

Yemen

Long Name: Republic of Yemen

Location: Middle East, bordering the Arabian Sea, Gulf of Aden, and Red Sea, between Oman and Saudi Arabia

Capital: Sanaa

Government Type: republic

Currency: Yemeni rial (YRl)

Area: 203,796 sq. mi. (527,970 sq. km)

Population: 16,388,000

Population Growth Rate: 3.3%

Birth Rate (per 1000): 43

Death Rate (per 1000): 10

Life Expectancy (from birth): 59.5

Total Fertility Rate (average children per woman): 7.1

Religion: Muslim including Sha'fi (Sunni) and Zaydi (Shi'a); small numbers of Jewish, Christian, and Hindu

Languages: Arabic

Literacy: 38%

GDP Per Capita: $2,900

Occupations: agriculture and service

Climate: mostly desert; hot and humid along west coast; temperate in western mountains affected by seasonal monsoon; extraordinarily hot, dry, harsh desert in east

Terrain: narrow coastal plain backed by flat-topped hills and rugged mountains; dissected upland desert plains in center slope into the desert interior of the Arabian Peninsula

433

Zambia

Long Name: Republic of Zambia

Location: Southern Africa, east of Angola

Capital: Lusaka

Government Type: republic

Currency: Zambian kwacha (ZK)

Area: 285,918 sq. mi. (740,720 sq. km)

Population: 9,461,000

Population Growth Rate: 2.1%

Birth Rate (per 1000): 45

Death Rate (per 1000): 23

Life Expectancy (from birth): 37.1

Total Fertility Rate (average children per woman): 6.4

Religion: Christian 50%-75%, Muslim and Hindu 24%-49%, indigenous beliefs 1%

Languages: English (official), Bemba, Kaonda, Lozi, Lunda, Luvale, Nyanja, Tonga, and about 70 other indigenous languages

Literacy: 78.2%

GDP Per Capita: $1,060

Occupations: agriculture 85%, mining, manufacturing, and construction 6%, transport and services 9%

Climate: tropical; modified by altitude; rainy season (October to April)

Terrain: mostly high plateau with some hills and mountains

Zimbabwe

Long Name: Republic of Zimbabwe

Location: Southern Africa, northeast of Botswana

Capital: Harare

Government Type: parliamentary democracy

Currency: Zimbabwean dollar (Z$)

Area: 149,255 sq. mi. (386,670 sq. km)

Population: 11,044,000

Population Growth Rate: 1.1%

Birth Rate (per 1000): 31

Death Rate (per 1000): 20

Life Expectancy (from birth): 39.2

Total Fertility Rate (average children per woman): 3.9

Religion: syncretic (part Christian, part indigenous beliefs) 50%, Christian 25%, indigenous beliefs 24%, Muslim and other 1%

Languages: English (official), Shona, Sindebele (the language of the Ndebele, sometimes called Ndebele), numerous but minor tribal dialects

Literacy: 85%

GDP Per Capita: $2,340

Occupations: agriculture 70%, transport and services 22%, industry 8%

Climate: tropical; moderated by altitude; rainy season (November to March)

Terrain: mostly high plateau with higher central plateau (high veld); mountains in east

Further Reading

Essential References

Clapson, D., C. Day, and J. Edwards (eds.) *Doring Kindersley World Reference Atlas.* Doring Kindersley Publishing Inc., 1994.

Espenshade, Edward B. Jr. (ed.) *Goode's World Atlas.* Rand McNally, 1995.

Famighetti, Robert. *World Almanac and Book of Facts.* St. Martin's Press, 1998.

MerriamWebster's Geographical Dictionary. MerriamWebster, 1997.

Microsoft Encarta Virtual Globe: 1998 Edition. CDROM. Microsoft Corporation, 1997.

Rand McNally Road Atlas: United States, Canada, Mexico. Rand McNally (annual).

Statistical Abstract of the United States. U.S. Department of Commerce (annual).

Web References

CIA World Factbook. http://geography.miningco.com/library/cia/blcindex.htm (annual)

Federal Emergency Management Agency. http://www.fema.gov

Library of Congress Country Studies. http://lcweb2.loc.gov/frd/cs/cshome.html

Rosenberg, Matthew T. *Geography at the Mining Co.* http://geography.miningco.com

United States Census Bureau. http://www.census.gov

General Geography

Barraclough, Geoffrey. *The Times Atlas of World History.* Hammond Incorporated, 1992.

Demko, George J. *Why in the World: Adventures in Geography.* Anchor Books, 1992.

Dickson, Paul. *Labels for Locals: What to Call People from Abilene to Zimbabwe.* Merriam-Webster, 1997.

Dunbar, Gary S. (ed.) *Modern Geography: An Encyclopedic Survey.* Garland Publishing, Inc., 1991.

Hirsch, E. D. Jr., Joseph F. Kett, and James Trefil. *The Dictionary of Cultural Literacy.* Houghton Mifflin Company, 1988.

Hudman, Lloyd E. and Richard H. Jackson. *Geography of Travel and Tourism.* Delmar Publishers, Inc., 1994.

Kenzer, Martin S. (ed.) *On Becoming a Professional Geographer.* Merrill Publishing Company, 1989.

Mayhew, Susan. *A Dictionary of Geography.* Oxford University Press, 1997.

Martin, Geoffrey J. and Preston E. James. *All Possible Worlds: A History of Geographic Ideas.* John Wiley & Sons, Inc., 1993.

Schlessinger, Arthur M. *Chronicle of the 20th Century.* DK Publishing, 1996.

Urdang, Laurence. *Names & Nicknames of Places & Things.* Meridian, 1987.

Cartography

Campbell, John. *Map Use and Analysis.* Wm. C. Brown Publishers, 1991.

Monmonier, Mark. *How to Lie With Maps.* University of Chicago Press, 1996.

Monmonier, Mark. *Cartographies of Danger: Mapping Hazards in America.* University of Chicago Press, 1997.

Wilford, John Noble. *The Mapmakers.* Alfred A. Knopf, 1981.

Physical Geography

Bolt, Bruce. *Earthquakes.* W.H. Freeman and Company, 1993.

DeBlij, H. J. *Physical Geography of the Global Environment.* John Wiley & Sons, Inc., 1996.

Lutgens, Frederick K. and Edward J. Tarbuck. *The Atmosphere: An Introduction to Meteorology.* Prentice Hall, 1995.

McKnight, Tom L. *Physical Geography: A Landscape Appreciation.* Prentice Hall, 1996.

Summerfield, Michael. *Global Geomorphology.* Longman Scientific & Technical, 1991.

Cultural, Urban, Transportation, and Political Geography

DeBlij, H. J. *Human Geography: Culture, Society, and Space.* John Wiley & Sons, Inc., 1993.

Jordan, Terry G. and Lester Rowntree. *The Human Mosaic: A Thematic Introduction to Cultural Geography.* Harper & Row, 1990.

Lewis, Tom. *Divided Highways: Building the Interstate Highways, Transforming American Life.* Viking, 1997.

Peters, Gary L. and Robert P. Larkin. *Population Geography: Problems, Concepts, and Prospects.* Kendall/Hunt Publishing Co., 1996.

Vance, James E. *Capturing the Horizon: The Historical Geography of Transportation.* Harper & Row Publishers, 1986.

Regional Geography

Birdsall, Stephen S. and John W. Florin. *Regional Landscapes of the United States.* John Wiley & Sons, Inc., 1991.

Blouet, Brian W. and Olwyn M. Blouet. *Latin America and the Caribbean: A Systematic and Regional Survey.* John Wiley & Sons, Inc., 1996.

DeBlij, H. J. and Peter O. Muller. *Geography: Realms, Regions, and Concepts.* John Wiley & Sons, Inc., 1997.

Craig, Gordon A. *Europe Since 1815.* Holt, Rinehart and Winston, 1966.

Glassner, Martin Ira. *Political Geography.* John Wiley & Sons, Inc., 1995.

Grove, A. T. *The Changing Geography of Africa.* Oxford University Press, 1989.

Hoffman, George. (ed.) *Europe in the 1990s.* John Wiley & Sons, Inc., 1990.

James, Preston. *Latin America.* The Odyssey Press, 1969.

Wheeler, Jesse H. and J. Trenton Kostbade. *World Regional Geography.* Saunders College Publishing, 1989.

Zinn, Howard. *A People's History of the United States.* Harper Perennial, 1980.

Exploration and Time

Boorstin, Daniel J. *The Discoverers.* Random House, 1983.

Doane, Doris. *Time Changes in the USA.* American Federation of Astrologers, 1994.

Fernandez Armesto, Felipe, (ed.) *The Times Atlas of World Exploration.* Harper Collins Publishers, 1991.

Pennington, Piers. *The Great Explorers.* Facts on File, 1979.

Index

KNOW IT ALL?
NOW YOU CAN!

The Handy Science Answer Book®

Can any bird fly upside down? Is white gold really gold? This best-selling book covers hundreds of new sci-tech topics from the inner workings of the human body to outer space and from math and computers to planes, trains and automobiles. *Handy Science* provides nearly 1,400 answers compiled from the ready-reference files of the Science and Technology Department of the Carnegie Library of Pittsburgh. Includes more than 100 illustrations.

1997 • Paperback • 598 pp.
ISBN 0-7876-1013-5

The Handy Weather Answer Book®

What is the difference between sleet and freezing rain? Do mobile homes attract tornadoes? You'll find clear-cut answers to 1,000 frequently asked questions in *The Handy Weather Answer Book*. A cornucopia of weather facts, *Handy Weather* covers such confounding and pertinent topics as tornadoes and hurricanes, thunder and lightning, and droughts and flash floods, plus fascinating weather-related phenomena such as El Niño, La Nina and the greenhouse effect. Includes 75 photos plus tables.

Walter A. Lyons, Ph.D. • 1996 • Paperback • 430 pp.
ISBN 0-7876-1034-8

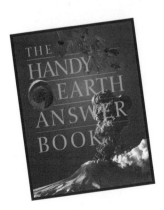

The Handy Earth Answer Book™

The Earth is the world's biggest celebrity. Natural disasters ... weather ... global warming ... and man's curious relationship with highest mountains, deepest oceans and extremes of heat and cold are always in the news. To satisfy everyone's earthly desires, there's *The Handy Earth Answer Book*. This easy-to-use, easy-to-read reference will entertain people of all ages. Topics covered in the book include physical geology, mountains, plate tectonics, geophysics, oceanography and more. Illustrated with 100 black and white drawings plus 40 color photos.

John Ernissee, Ph.D./Frank Vento, Ph.D. • 1999 • Paperback • 425 pp.
ISBN 1-57859-050-7